Achievements of the National Plant Genome Initiative and New Horizons in Plant Biology

Committee on the National Plant Genome Initiative:
Achievements and Future Directions

Board on Life Sciences

Board on Agriculture and Natural Resources

Division on Earth and Life Studies

NATIONAL RESEARCH COUNCIL
OF THE NATIONAL ACADEMIES

THE NATIONAL ACADEMIES PRESS
Washington, D.C.
www.nap.edu

THE NATIONAL ACADEMIES PRESS 500 Fifth Street, N.W. Washington, DC 20001

NOTICE: The project that is the subject of this report was approved by the Governing Board of the National Research Council, whose members are drawn from the councils of the National Academy of Sciences, the National Academy of Engineering, and the Institute of Medicine. The members of the committee responsible for the report were chosen for their special competences and with regard for appropriate balance.

This study was supported by the Interagency Working Group on Plant Genomes through Contract/Grant No. DBI-0722206 between the National Academy of Sciences the National Science Foundation and the Department of Energy, Contract/Grant No. 59-0206-7-140 from the U.S. Department of Agriculture (USDA) Agricultural Research Service (ARS), Contract/Grant No. 2007-35300-17910 from USDA Cooperative State Research, Education, and Extension Service (CSREES), and Contract/Grant No. 07-DG-11132650-286 from the USDA Forest Service (USFS). The content of this publication does not necessarily reflect the views or policies of the sponsoring agencies, nor does the mention of trade names, commercial products, or organizations imply endorsement by the U.S. Government.

International Standard Book Number-13: 978-0-309-11418-9 (Book)
International Standard Book Number-10: 0-309-11418-7 (Book)
International Standard Book Number-13: 978-0-309-11419-6 (PDF)
International Standard Book Number-13: 978-0-309-11419-5 (PDF)

Libary of Congress Control Number: 2008923010

Additional copies of this report are available from the National Academies Press, 500 Fifth Street, NW, Lockbox 285, Washington, DC 20055; (800) 624-6242 or (202) 334-3313 (in the Washington metropolitan area); Internet, http://www.nap.edu.

Cover: Design by Michele de la Menardière.
Photo credits:
Foreground on the left: Inhibition of RNA silencing pathways in *Arabidopsis* as shown by false-colored scanning electron micrographs of early-stage flowers from wild-type or silencing-defective plants. Blue flowers are the wild-type. This photo was published on the cover of *Developmental Cell,* Vol. 4, Issue 2, Copyright Elsevier (2003). Reprinted with permission.
Background on the left: Rice field taken during the "Rice Research to Production" course at the International Rice Research Institute, Los Banos, Philippines in June 2007. Copyright Michael J. Kovach (2007). Reprinted with permission.
Right: Soybeans ready for harvest. Photo by Scott Bauer, ARS Photo Unit.

Printed in the United States of America.

THE NATIONAL ACADEMIES

Advisers to the Nation on Science, Engineering, and Medicine

The **National Academy of Sciences** is a private, nonprofit, self-perpetuating society of distinguished scholars engaged in scientific and engineering research, dedicated to the furtherance of science and technology and to their use for the general welfare. Upon the authority of the charter granted to it by the Congress in 1863, the Academy has a mandate that requires it to advise the federal government on scientific and technical matters. Dr. Ralph J. Cicerone is president of the National Academy of Sciences.

The **National Academy of Engineering** was established in 1964, under the charter of the National Academy of Sciences, as a parallel organization of outstanding engineers. It is autonomous in its administration and in the selection of its members, sharing with the National Academy of Sciences the responsibility for advising the federal government. The National Academy of Engineering also sponsors engineering programs aimed at meeting national needs, encourages education and research, and recognizes the superior achievements of engineers. Dr. Charles M. Vest is president of the National Academy of Engineering.

The **Institute of Medicine** was established in 1970 by the National Academy of Sciences to secure the services of eminent members of appropriate professions in the examination of policy matters pertaining to the health of the public. The Institute acts under the responsibility given to the National Academy of Sciences by its congressional charter to be an adviser to the federal government and, upon its own initiative, to identify issues of medical care, research, and education. Dr. Harvey V. Fineberg is president of the Institute of Medicine.

The **National Research Council** was organized by the National Academy of Sciences in 1916 to associate the broad community of science and technology with the Academy's purposes of furthering knowledge and advising the federal government. Functioning in accordance with general policies determined by the Academy, the Council has become the principal operating agency of both the National Academy of Sciences and the National Academy of Engineering in providing services to the government, the public, and the scientific and engineering communities. The Council is administered jointly by both Academies and the Institute of Medicine. Dr. Ralph J. Cicerone and Dr. Charles M. Vest are chair and vice chair, respectively, of the National Research Council.

www.national-academies.org

COMMITTEE ON THE NATIONAL PLANT GENOME INITIATIVE: ACHIEVEMENTS AND FUTURE DIRECTIONS

JEFFERY L. DANGL, *Chair,* University of North Carolina, Chapel Hill
LOIS BANTA, Williams College, Williamstown, MA
ROGER BOERMA, University of Georgia, Athens
JAMES C. CARRINGTON, Oregon State University, Corvallis
JOANNE CHORY, Salk Institute for Biological Studies, San Diego, CA
STEVE A. KAY, University of California, San Diego
SUZANNA LEWIS, Lawrence Berkeley National Laboratory, Berkeley, CA
THOMAS MITCHELL-OLDS, Duke University, Durham, NC
NEELIMA R. SINHA, University of California, Davis
MICHAEL SNYDER, Yale University, New Haven, CT
STEVEN H. STRAUSS, Oregon State University, Corvallis
ERIC R. WARD, Two Blades Foundation, Durham, NC

Staff

EVONNE TANG, Study Director
FRANCES E. SHARPLES, Director, Board on Life Sciences
ROBIN SCHOEN, Director, Board on Agriculture and Natural Resources
MERC FOX, Program Assistant
PAULA WHITACRE, Editor

BOARD ON AGRICULTURE AND NATURAL RESOURCES

W. REG GOMES, *Chair,* University of California, Agriculture and Natural Resources (*Emeritus*)
SANDRA J. BARTHOLMEY, University of Illinois, Chicago
ROGER N. BEACHY, Donald Danforth Plant Science Center, St. Louis, MO
H.H. CHENG, University of Minnesota (*Emeritus*), St. Paul
DANIEL M. DOOLEY, Dooley Herr & Peltzer, LLP, Visalia, CA
JOAN H. EISEMANN, North Carolina State University, Raleigh
BRUCE L. GARDNER, University of Maryland, College Park
HANS R. HERREN, Millennium Institute, Arlington, VA
KIRK C. KLASING, University of California, Davis
VICTOR L. LECHTENBERG, Purdue University, West Lafayette, IN
BRIAN W. MCBRIDE, University of Guelph, Ontario, Canada
ROBERT PAARLBERG, Wellesley College, Watertown, MA
BOBBY PHILLS, Florida A&M University, Tallahassee
KEITH PITTS, Pew Initiative on Food and Biotechnology, Fair Oaks, CA
SONYA SALAMON, University of Illinois, Urbana-Champaign (*Emeritus*)
HAL SALWASSER, Oregon State University, Corvallis
PEDRO A. SANCHEZ, The Earth Institute at Columbia University, Palisades, NY
NORMAN R. SCOTT, Cornell University, Ithaca, NY
BILLIE L. TURNER II, Clark University, Worcester, MA
LAURIAN J. UNNEVEHR, University of Illinois, Urbana-Champaign
TILAHUN D. YILMA, University of California, Davis

Staff

ROBIN SCHOEN, Director
RUTH S. ARIETI, Senior Program Assistant
KAREN L. IMHOF, Administrative Assistant
AUSTIN J. LEWIS, Program Officer
MICHAEL MA, Visiting Program Officer
MARGOT RHYU, Program Assistant
JANET MULLIGAN, Research Associate
EVONNE TANG, Senior Program Officer
PEGGY TSAI, Associate Program Officer

Preface

Why should the public and the government care about plant science, let alone the proposition of a National Plant Genome Initiative? Well, quite simply, human life would be impossible without plants. You are likely reading this document wearing plant-derived clothing (cotton, or wool from a sheep that ate a plant before shearing), sitting in or on a plant-derived product (wood), and digesting your last meal, made up largely of plants, or something that ate a plant as its last meal before you ate it! And you probably got to work today using transportation that runs on fossil fuel made from plants that lived a couple of hundred million years ago. Without plants, we'd be in serious trouble. Photosynthesis and carbon sequestration are performed by organisms of all sizes and shapes, from tiny oceanic phytoplankton and single-celled alga, to giant redwood and teak trees, to dense assemblages of diverse plants in rainforest and grassland communities around the planet. Plants and microorganisms, and the communities they form, are fundamental contributors to, and regulators of, the entire earth system.

Modern plant biology, then, is about human health and well being, nutrition from new and better foods, fiber and wood production, and renewable alternatives to imported fossil fuels, all of which are produced under the umbrella of responsible environmental stewardship. These issues have been focused by the recent broad acceptance that global climate change is upon us, and that human activity contributes to at least some of it. Hence, the imperative to understand plants, how they grow and how they produce the products upon which we all depend, has never been more immediate. Plants are a cornerstone to all animal life, including humans.

Yet, because we in the United States are no longer an agrarian society, but rather an urban or suburban one with food and fiber surpluses, many of us have lost touch with the simple conceptual notion that plants and plant products remain vital to the way we live. It is also a simple, but profound, truth that plants and plant communities add beauty, flavor, fragrance, and tranquility to our existence. Despite these facts, we, as a society, do not typically consider our reliance on plants, and plant biology, as an important linchpin in U.S. technological and research infrastructure or in our national security, in the same way that we consider public investment in human health research a critical societal priority. For example, the Human Genome Project and the promise of treatments and therapies derived from it have captured the attention of most everyone who watches the news or reads a newspaper.

That situation is changing because of the public's concern with climate change and the recent public interest in alternative and renewable energy sources, including biofuels. But very few of our constituents make the link between plant genome research and the potential to help generate solutions to the very challenging societal problems that we face. We, as scientists and science policy makers, can do a better job getting that message out. The popular success of books with plant biology and food production as their theme suggests that there is a public interested in plant biology writ large. And, most importantly, our message is compelling on economic competitiveness grounds, on geopolitical grounds, on scientific grounds, and on aesthetic grounds.

The National Plant Genome Initiative (NPGI) represents a unique, cross-agency funding enterprise for plant genomics coordinated by the Interagency Working Group (IWG), which is comprised of various federal agencies. As the NPGI approaches its 10-year birthday, the IWG asked the National Research Council (NRC) to evaluate the program and to make recommendations as to how the future of plant genome science might look. The NRC's Board on Life Sciences and the Board on Agriculture and Natural Resources convened a committee of experts in the fields of plant genetics, epigenetics, informatics, biology education, and molecular and cell biology to evaluate the NPGI program, which involved reviewing the science produced, the science soon to come, and the science that the most creative plant biologists in the country could envision as accomplishments for the NRC report that will be written 20 years from now. The committee did this in the form of a meeting with the IWG, a community workshop, a short questionnaire sent to the more than 270 principal investigators (PIs) who have been funded by NPGI over its first nine years, and a long series of telephone conferences. I would like to thank all the committee members, all the PIs who responded to our questionnaire, and all the participants of the meeting and workshop. We hope that your opinions and advice have been used wisely in this report.

NPGI is intertwined unofficially with the National Science Foundation's other flagship plant biology activity, the *Arabidopsis* 2010 Project, which focuses on using a powerful model species to understand many of the basic conceptual aspects of plant growth and development that provide touchstones across at least the flowering plants. Hence, NPGI and *Arabidopsis* 2010 are mentioned together at many junctures in this report. That's good, and should raise no particular concerns among readers, since the two programs are very complementary and truly synergistic. This is easily illustrated in my own discipline, the study of the plant immune system. Over the last 15 years, genetics-based research using *Arabidopsis* has led to a fairly detailed wiring diagram of the plant immune network. But that diagram benefited from seminal discoveries using flax, tomato, barley, rice, and tobacco. In turn, the conceptual generalities defined in *Arabidopsis* have subsequently been shown to operate in many other species.

Modern molecular, cellular, and developmental biology is the story of the adoption of easily manipulated model organisms that serve to provide the "big picture" for a much broader set of scientific truths. Thus, the classic case of research using lab mice and fruit flies that, while of course very compelling in its own right to those scientists who do the work, is easily tied to arguments equating model organism research with breakthroughs in human health. In other words, basic science using animal models has as its ultimate goal explicit betterment of human health. This case is intrinsically more difficult to make for plant science, since we, as humans, use many different plant species for different purposes. The tricky policy issue is how to inform the improvement of many of these plant species using genomics with what will always be limited resources.

The multiagency IWG structure of NPGI presented the committee with some problems. For example, it was initially difficult to parse out what is actually funded by NPGI. But the IWG participants in the NPGI worked closely and patiently with the committee over many months to ensure that we had access to the correct sets of data necessary to review both the science and the unique research management concept embodied in the IWG.

I hope that this report generates discussion in the research, education, and policy communities. The committee feels strongly that a broad examination of the accomplishments of NPGI and *Arabidopsis* 2010 Project will be rewarding, and will easily justify rapid growth in federal investment in plant genomics. The case for an expansion of the program is compelling—the infrastructure built over the first nine years of NPGI is in place and now maturing, and the challenges to society that have their solutions in plant science have never been greater.

Jeffery L. Dangl
Chair, Committee on the National Plant Genome Initiative: Achievements and Future Directions

Acknowledgments

This report is a product of the cooperation and contributions of many people. The members of the committee thank all of the speakers who briefed the committee. (Appendix D presents a list of presentations to the committee.)

This report has been reviewed in draft form by persons chosen for their diverse perspectives and technical expertise in accordance with procedures approved by the National Research Council's Report Review Committee. The purpose of this independent review is to provide candid and critical comments that will assist the institution in making its published report as sound as possible and to ensure that the report meets institutional standards of objectivity, evidence, and responsiveness to the study charge. The review comments and draft manuscript remain confidential to protect the integrity of the deliberative process. We wish to thank the following for their review of this report:

James Birchler, University of Missouri
Sean Eddy, Janelia Research Farm
Michael Herman, Kansas State University
Amy Iezzoni, Michigan State University
Jan Leach, Colorado State University
Michael Lynch, Indiana University
Elliot Meyerowitz, California Institute of Technology
Ronald Phillips, University of Minnesota
Gordon Uno, University of Oklahoma
Detlef Weigel, Max Planck Institute for Developmental Biology

Although the reviewers listed above provided constructive comments and suggestions, they were not asked to endorse the conclusions or recommendations, nor did they see the final draft of the report before its release. The review of this report was overseen by Dr. R. James Cook, Washington State University. Appointed by the National Research Council, Dr. Cook was responsible for making certain that an independent examination of this report was carried out in accordance with institutional procedures and that all review comments were carefully considered. Responsibility for the final content of this report rests entirely with the author committee and the institution.

Contents

APPENDIXES

Summary

Humans have used plants as major sources of food, fiber, energy, and animal feed for millennia. The organized harnessing of plant products through domestication was a key step in the ascendance of human civilization. Subsequent modern plant breeding has led to greatly enhanced productivity and incorporation of new, desirable traits into a wide variety of plants used by humans for diverse applications. Despite the long history of human-plant interactions, investment in plant biology research has never been as important as it is today, as a growing global human population increases the demand on plant production in the face of decreased fresh water supplies and societal pressures to maintain the quality of arable and "natural" land. The need to enhance plant production will be exacerbated by extreme weather patterns as a result of global climate change.

Plant genome science facilitates the integrated study of complex, important traits with value to human society. Plant biomass productivity, chemical composition, grain and fruit yield, adaptability to suboptimal environments, and defensive responses to pests are genetically conditioned traits. All derive from the integrated contributions of multiple genetic networks. In fact, the principle that plant performance traits are the result of complex genetic determinants acting under various environmental influences underlies most current strategies for plant improvement. The same principle also predicts that gains in plant productivity will be best achieved through tools for systematic analysis and genome characterization that are enabled by plant genome sciences.

The National Plant Genome Initiative (NPGI) was initiated in 1998 in

BOX S-1
Charge to the Committee

- Review the accomplishments of NPGI to date.
- Assess the contribution of NPGI to science, research infrastructure, education of the next generation of scientists, and international research collaboration.
- Discuss the broad impacts of NPGI to fundamental advances in biological sciences.
- Assess the contributions of NPGI to the application of scientific knowledge including technological innovation and economic competitiveness.
- Recommend future research directions and objectives for NPGI.

recognition that research in plant genomics provides a foundation for rapid, fundamental, and novel insights into the means by which plants grow and reproduce, produce organs and tissues essential to human nutrition and energy production, adapt to different and sometimes stressful environments, and help stabilize ecosystems. As NPGI approaches its 10th anniversary, the Interagency Working Group on Plant Genomes[1] (IWG) asked the National Research Council (NRC) to convene a committee to assess the achievements of NPGI and recommend future research directions (Box S-1). To address the statement of task, the committee gathered information from IWG, from principal investigators who received research grants from NPGI, and from plant scientists from universities, government agencies, and industry through a workshop to solicit their evaluation of the achievements of NPGI in the last nine years and to discuss possible future directions of the program.

The assessment of NPGI comes at a vital period in the history of federally supported scientific research. Budgets are more or less flat or, in some cases, declining. Yet the needs for revolutionary breakthroughs and technological advancement are acute and international competition is increasing, so that a new vision for plant genome sciences over both the short and long term is necessary.

[1]The Interagency Working Group on Plant Genomes (IWG) was established in May 1997 by the Office of Science and Technology Policy (OSTP) under the direction of the National Science and Technology Council's Committee on Science, in recognition of the unprecedented scientific opportunities that plant genome research offered at that time. IWG membership currently includes the National Science Foundation, U.S. Department of Agriculture, U.S. Department of Energy, National Institutes of Health, U.S. Agency for International Development, U.S. Forest Service, Office of Management and Budget, and OSTP.

ASSESSMENT

NPGI has been very successful by all measures applied in this study. NPGI has contributed to revolutionary breakthroughs in genome sequencing of plants, including the genomes of *Arabidopsis*, rice, and, soon, maize and a variety of these plants' pests and pathogens. Genome sequence, however, is only the first step toward understanding integrated gene function. NPGI has supported a wide range of investigations into gene regulatory mechanisms in model and crop plants and funded studies on natural variation and crop domestication to better understand complex traits. Furthermore, NPGI has enabled the development of more efficient DNA-based breeding tools to improve crop value and yield. The NPGI funding has also supported the development of artificial chromosome libraries for several crops and research to align these DNA sequences in their proper location along the normal chromosomes. **Far more than just genomics, the technologies and information developed by NPGI and by the parallel and complementary program *Arabidopsis* 2010 Project of the National Science Foundation (NSF)[2] are the primary platforms for *basic research* in fundamental plant science—including genetics, biochemistry, physiology, developmental biology, evolutionary biology, and population biology.**

Plant genome scientists, as a community, have made excellent use of the synergies provided by exploitation of an easily manipulable model species, *Arabidopsis*, to elucidate basic biological principles that are likely to be broadly operative across plant biology and can thus facilitate rapid applications to crop genomics and improvement. In fact, it is difficult to list all the major breakthroughs that the analyses of plant genomes and gene function have enabled in the last nine years across NPGI and the NSF *Arabidopsis* 2010 Project. These have been powered mostly by genetic advances in *Arabidopsis*, and have often been followed up by studies of similar genes and their related function in crop species. Among them are the discovery of receptors for nearly all of the major plant hormones; an increasingly detailed understanding of how these receptors control subsequent plant developmental programs; knowledge of how exposure to winter-like temperature, vernalization, and the correct photoperiod leads to flowering; how flowers, leaves, and roots are built; and how the plant "immune system" controls the interplay of the different lines of pathogen defense. Some of these breakthroughs are now being translated to practical applications in crop species. **This synergism**

[2]The NSF's *Arabidopsis* 2010 Project is not part of the committee's review, but it is unofficially intertwined with NPGI. The *Arabidopsis* 2010 Project focuses on using a rapid growing model species to understand many of the basic aspects of plant growth and development that provide conceptual touchstones across at least the flowering plants (NSF 2007a).

is the best rationale for further and separate development of both NPGI and the independently funded *Arabidopsis* 2010 Project.

Additionally, the direct application of genomics to less tractable, but economically important, plant species and to less intensively studied ecological models is increasingly feasible, particularly as costs for many genomics technologies decline. These genomic studies have led to recent breakthroughs in the understanding of the molecular events controlling the traits that were selected by humans during domestication of, for example, maize, rice, tomato, and wheat. Similarly, genomics is now being used to understand evolutionary processes in ecologically relevant settings and guide breeders toward new forms of crop adaptation to stressful environments.

The significance of NPGI-funded research is reflected by the impact of the articles published by the principal investigators. The NPGI "literature footprint" is heavily weighted toward important, broadly applicable publications in the best scientific journals. About 21 percent of primary peer-reviewed research articles that cited NPGI support were published in journals with citation impact factor of 9 or above, and around 45 percent of them were published in journals with citation impact factor of 6 or above; this is an excellent record. In addition, there have been many publications in crop breeding and agronomic journals, demonstrating transfer of technology to the applications sector. The publication record shows that NPGI research is not only a leading source of plant biology knowledge, but also a source of research that is broadly applicable to the wider biosciences community.

Basic research funded by NPGI to date has served as the springboard for several applied, agency-specific, mission-oriented programs that capitalize on either new funding from the public or on public-private partnerships. Examples of programs that leverage basic plant genomics discoveries made though NPGI include the U.S. Department of Energy (DOE) Bioenergy Research Centers and the Plant Feedstocks Genomics Bioenergy Program run jointly by DOE and the U.S. Department of Agriculture (USDA) Cooperative State Research, Education and Extension Service (CSREES). The USDA Agricultural Research Service and CSREES have refocused or increased investment in their agency-specific programs as a result of NPGI discoveries. The committee interprets these agency-specific research programs in plants to be a function of both maturation of NPGI and a broadly accepted view that plant biology will contribute significantly to long-term energy solutions.

NPGI principal investigators also reported diverse and substantive translational activities. These activities range from starting their own companies on the basis of research results to patent filings and licensing arrangements with a variety of plant biotechnology entities. Hence, NPGI research is moving towards real world application and crop improvement.

In addition to advancing basic science, NPGI has contributed to international collaborative projects and to the training of a large number of students and postdoctoral fellows. These efforts have created a pool of employees for growing enterprises in all bioscience sectors in the United States and around the world. A number of NPGI-funded projects involve international partners, and these projects leverage the resources, expertise, and facilities of many countries to achieve a richer and more comprehensive set of genome datasets than could be obtained by any single national effort. These collaborations can enhance the competitive ability of the United States to contribute to solutions to global challenges. For example, coordination among researchers from six groups across three continents in the *Arabidopsis* genome sequencing project paved the way for subsequent multinational endeavors, including the international consortium to sequence the rice genome. NPGI is a model for international scientific collaboration, and the ability of the United States to contribute to global challenges has been significantly enriched by these research programs.

NPGI is not merely a funding mechanism, but it also is an interagency collaboration that coordinates activities in plant genomics. All of the member agencies contribute to the goals of NPGI by providing in-kind support, distributing resources, and keeping each other abreast of the latest genomic technologies to ensure efficient use of resources. Other than the *Arabidopsis* 2010 Project and the USDA-CSREES National Research Initiative, there are few other funding sources in the United States that support the basic disciplines of plant science that are comparable to NPGI.

The committee therefore takes it as axiomatic throughout this report that the crop-focused NPGI and the independently funded, rapid discovery *Arabidopsis* 2010 Project should continue in parallel. To miss the opportunity to capture and increase the momentum of the last 10 years would diffuse the tremendous gains made thus far, and sacrifice vital routes to tackle national and global problems that could be addressed with plant-based resources.

RECOMMENDATIONS AND GOALS: NEW HORIZONS IN PLANT GENOMICS

The committee's major recommendations incorporate a significant broadening of the NPGI mission to include the basic biology of economically relevant traits in models and crop species, deeper investigations into plant diversity and plant adaptation to various ecological niches, and continued expansion of translation to breeders and farmers. This broadening is justified by the wealth of knowledge to be gained from comparative genomic analyses within and across species, and by the need to understand how plants function to provide, in essence, the conditions required for human survival. The diversity of plants, combined with hundreds of millions of years of natural selection, has resulted in evolutionary "solutions" to problems of growth and survival across many radically different ecological conditions, and in the face of large variations in environmental stress. The conserved nature of genetic networks across species and the ability to transfer knowledge from one species to another via comparative genomics and subsequent marker-assisted breeding, and by direct genetic engineering in many cases, can lead to potential major innovations in crop improvement.

The committee's recommendations for an expanded NPGI will require research at scales ranging from single principal investigators, across collaborations of investigators at multiple institutions, to large dedicated data production centers, all of which will use experimental methods of inquiry that span the entire scale of plant biology. The scale ranges from atomic-level analysis of plant cell constituents to analyses of agronomic yield and plant organ or tissue composition, and from the study of individual cells to population-wide analyses of whole-plant phenotypes in field environments or natural ecological communities. Indeed, for effective translation of the insights from plant genomics to aid crop improvement, conservation, and ecological studies, considerable efforts need to be allocated to studying plant phenotypes and physiology in both realistic agricultural and natural environments. The astounding breadth of plant genomics will attract, motivate, and empower scientists across a wide spectrum of disciplines and should result in synergistic gains in knowledge and application.

The committee focused its recommendations on three different time horizons: The 5-year goals represent immediate, pragmatic "next steps" in plant genome science, the 10-year goals require significant development of new tools and resources to enable transformative solutions to real world problems, and the 20-year "achievements" reflect the committee's desire to define some admittedly long-range, high-risk, high-reward areas that would significantly alter society's ability to understand how plants work. The committee used the following guiding principles for the development of the recommendations across these time horizons:

- NPGI research should encompass the most innovative, competitive, peer-reviewed basic science aimed at detailed and system-wide understanding of plant form, function, performance, and evolution in a strictly peer-reviewed, competitive environment.
- Addressing core molecular, cellular, and developmental concepts of plant biology is most efficient and productive in highly developed model plant systems; research most easily done in these systems should receive high priority.
- The diversity of plant form and function utilized by humans is very broad; hence *parts* of the overall genomics toolkit should be devoted to investigate spe-

cific aspects of plant tissue and organ development, environmental adaptations, or biochemical processes that are not well represented in core model species.

- Field-robust, high-resolution genotyping and phenotyping methods for use in studies of natural adaptations and in molecular-assisted plant breeding should be available across a broad swath of wild and cultivated plant species.
- Priorities for NPGI and associated plant sciences should be framed towards addressing the large challenges facing humanity, including biobased energy, climate change, sustainability, food security, and human health and nutrition.

The committee's major recommendations are that NPGI should:

1. Expand plant genome sequencing, plant-associated microbial sequencing, and plant-associated metagenome sequencing, and associated high quality annotation, by (a) using the Department of Energy's Joint Genome Institute's sequencing capacity to generally serve plant sciences and (b) empowering individual principal investigators or collaborative groups to access and utilize next-generation sequencing technologies for a broad spectrum of genomics and metagenomics discovery.

2. Develop "omics" resources and toolkits at high resolution in a few, carefully chosen plant species, including expansion and deeper investment in currently leading model species.

3. Develop "omics" resources at a broader, shallower level across a number of additional species to (a) expand the phylogenetic scope of functional inference, particularly when this is justified to test clearly specified hypotheses, (b) understand physiological and developmental processes to a depth that is not feasible in the model systems, and (c) provide the foundation to improve U.S. competitiveness of important crop and tree species.

4. Use systems-level approaches to understand plant growth and development in controlled and relevant environments, with the goal to create the iPlant, a large family of mathematical models that generate computable plants genuinely predictive of plant system behavior under a range of environmental conditions.

5. Increase the understanding of plant evolution, domestication, and performance in various ecological settings via investment in comparative genomics, and in the metagenomics of living communities of interacting organisms.

6. Enable translation of basic plant genomics towards sustainable deliverables in the field, and continue to use NPGI as a foundation for new, agency-specific, mission-oriented plant improvement programs.

7. Develop and deploy sustainable, adaptable, interoperable, accessible, and evolvable computational tools to support and enhance Recommendations 1–6.

8. Improve the recruitment of the best, broadly trained scientists into plant sciences.

9. Promote outreach on plant genomics and related issues that are critical to educating the American public on the value of genomics-based innovations.

These recommendations are very well aligned with Federal Research and Development Budget Priorities (http://www.ostp.gov/html/budget/2008/m06-17.pdf) as articulated by Office of Science and Technology Policy and the Office of Management and Budget:

> Agencies should target investments toward the development of a deeper understanding of complex biological systems, which will require collaborations among physical, computational, behavioral, social, and biological scientists and engineers who will, among other things, need to develop the data management tools and platforms necessary to facilitate this research. Access to new biotechnological tools and increasing amounts of genetic sequence data will open new avenues for research into the functional implications of gene expression. . . . Continued research at both the cellular/sub-cellular and the organism/community levels has the potential to have significant impact on national security and homeland security, health, environmental management, and education.

CONCLUSION

Plant genome sciences, and plant biology as a whole, are vital enterprises that contribute significantly to human health, energy security, and enlightened and careful environmental stewardship. Because photosynthetic organisms play a central role in all of the Earth's major ecosystems, understanding how plants function, and how to modify and improve their ability to carry out specific physiological processes—the goals of plant genome research—are likely to have deep and far-reaching ultimate environmental impacts. NPGI demonstrates that the plant genome science research community is vibrant and capable of continued imaginative breakthroughs that will drive technological advancement. IWG should capitalize on the research capacity built and important knowledge gained via NPGI to date and expand the program so that the member organizations' current and future agency-specific and mission-oriented programs will always be built on a strong foundation of fundamental plant biology research.

Akin to President Kennedy's call to land an astronaut on the moon, or President Nixon's declaration of war on cancer, the time is now ripe for a major, national research effort toward sustainability in the production of food, fuel, and fiber in a societal climate that should demand responsible environmental stewardship. Our society's overall response to these challenges will entail conservation, rapid and significant gains in fuel use efficiency, and the development of a distributed, diversified energy portfolio. Plants will provide a necessary part of the solutions to several of these issues. Plant genome science research can play a large part in this important revolution.

1

Introduction

PLANT SCIENCES: VITAL TO HUMAN HEALTH AND EXISTENCE

Nearly all organisms on Earth, including humans, depend on plants for survival. Plants, along with photosynthetic marine algae, are the primary converters of solar energy into the usable, stored forms of energy that power life on Earth. Thus, current and future breakthroughs in plant biology research can have profound consequences for the future of humanity and for the entire biosphere. The rationale for expanded investments in *plant genome research* is straightforward and urgent: Plant genomics provides a foundation for rapid, fundamental, and novel insights into the means by which plants grow and reproduce, produce organs and tissues essential to human nutrition and energy production, adapt to different and often difficult environments, and help stabilize ecosystems. Plant genomics is already beginning to enable a variety of new technologies that will revolutionize plant breeding and enhance responsible stewardship of the environment.

The United States and the world face enormous challenges related to the production of food and energy, maintenance of environmental quality, and mitigation of climate change. The maintenance of food quantity, quality, nutritional content, and delivery are persistent issues, as reflected in chronic food crises around the world. To resolve the environmental problems derived from the growing use of nonrenewable fossil fuels, renewable but practical and environmentally sustainable alternatives are necessary.

Climate change will force adaptation in plant communities, which will result

in changes in plant distribution that are likely to have tremendous impact on human well being and ecosystem sustainability. Long-term records have indicated that the Earth's atmosphere is warming at an unprecedented rate (Trenberth et al. 2007). The impacts of climate change will likely be highly variable in space and time, leading to difficult-to-predict outcomes in different parts of the world. However, predicted effects include an increase in new outbreaks of pathogen and pest infestations, and an increased frequency of extreme climate events such as droughts, fires, and floods. These predicted effects could have severe impacts on agriculture and forestry (Easterling et al. 2007). Climate change as a result of asymmetries in CO_2 emissions and carbon sequestration, and growing water shortages are likely to lead to dramatic changes in agricultural productivity and land use and availability (Reddy and Hodges 1999). By increasing knowledge of how plants cope with extreme stresses, plant genomics research can help scientists to more precisely breed or engineer plants that can thrive as climates change.

Economically and energetically viable production of liquid fuels from plant biomass, in quantities that could contribute to a reversal in the world's dependence on fossil fuels, will require increases in plant productivity and concomitant advances in biomass-to-fuel conversion. Directed modification of plant productivity and the tailoring of lignocellulosic biomass for high rates of conversion to liquid fuels increasingly depends on plant genomics to describe, at high resolution, the pathways that control biomass production, structure, and chemistry (DOE 2006).

Sustainable agriculture will require a reduction in fossil fuel-derived inputs and in agriculturally caused pollution (for example, runoff of excess nitrogen, phosphorus, potassium, and various pesticides) and soil degradation (for example, loss of soil carbon, and associated fertility and soil loss as a result of erosion). Meeting these goals will depend, in part, on technological advances suited to a wide variety of agricultural and ecological conditions around the world. Most crops in most years are harvested at yields that are not nearly as high as their corresponding record yields (Boyer 1982). Hence, optimal plant performance, which depends on the convergence of weather, water, and soil conditions and subsequent genome-encoded physiological responses, is much higher than what is typically achieved. Plant genomics research can contribute to understanding the mechanisms that determine optimal plant performance by identifying natural mechanisms governing plant growth, development, and adaptation to weather and water stress, and by helping to catalogue the evolutionary diversity of agriculturally important genes.

Basic plant genome research serves a wide diversity of agricultural and environmental goals. Agricultural production in the United States can be broadly divided into three categories: large commodity crops (such as corn, soybean, wheat, sorghum, cotton, and forage species), specialty crops (including fruits, nuts, vegetables, and ornamental species), and forest trees. The traits of importance vary widely from emphasis on yield and on tolerance to stress in the commodities; to flavor, scent, and nutritional composition in the fruits and vegetables (Goff and Klee 2006); to color, form, shape, and pest tolerance in ornamentals; and to wood or fiber yield and quality in forest trees. All crops are produced across farms that use various economic models, ranging from large farms that use vertically integrated production and crop systems that focus on one or a few rotation species, to small acreage and diverse plantings in produce farms that sell directly to consumers through local farmers' markets. All these farming models can be served by investment in basic plant genome research.

Plant genome science facilitates the otherwise difficult integrated study of complex, economically important traits. Plant biomass productivity, chemical composition, grain and fruit yield, adaptability to suboptimal environments, and defensive responses to pests are genetically conditioned traits. All of them derive from the integrated contributions of multiple genetic networks. The principle that plant performance traits have complex genetic determinants underlies most current strategies for plant improvement, which emphasize phenotypic evaluation of whole plants in realistic environments. The same principle also predicts that gains in plant productivity will be best achieved through tools for systematic analysis and genome characterization that are enabled by plant genome sciences.

The fundamental goals of plant genome science are to understand plant growth, form, function, adaptation, diversity, and evolution. That knowledge is critical to sustained progress in plant improvement. Advances in basic plant biology not only help to direct breeding of current crops and traits of value, but also stimulate progress in new directions, such as domestication of new crops and generation of new types of crop products. Domestication of new crops is of special importance in the face of rapidly changing climates and global markets. Examples of how new plants and new uses for plants can arise include the emergence of soybean to become a dominant commodity from relative obscurity over the last 60 years, and the exploitation of kiwi fruit as a new specialty crop over the last few decades.

The breeding of major commodity crops already is benefiting directly from plant genomics research. The four largest commodity crops—corn, soybean, sorghum, and wheat—have an annual farm gate value (defined as value of crops to the grower) of $75 billion (NASS 2007a,b). The output of those crops is processed into a wide range of food, feed, and industrial applications, cutting a broad swath across the U.S. economy. Plant genomics efforts to date have focused, to a large extent, on these crops, driving an expansion of basic knowledge and establishment of research platforms, as well as the application of DNA markers to increase the efficiency of public and private sector breeding (see Chapter 2). In industry, the application of

DNA marker-assisted selection for pest resistance in soybeans has resulted in the cost-effective development of high-performing soybean cyst nematode-resistant cultivars (Calhill and Schmidt 2004).

The specialty crops as a group are economically valuable and are used to generate products with which the public is intimately familiar. Specialty crops, too, are poised to benefit from plant genomics research. Despite their limited acreage compared to the commodities, the total annual value of specialty crops is about $50 billion (ERS 2007). However, that value is highly segmented: For example, $10 billion of value is divided among 34 major types of vegetables (NASS 2007c). Research aimed at specialty crop improvement is confounded by the market segmentation; some crops receive relatively high levels of attention from public and private breeders, and others little or none. The paucity of basic plant genomic information and the cost of its application are current limiting factors to the improvement of nearly all specialty crops. Plant genomics research in the public sector and associated DNA sequence databases, and the rapid decline in costs of sequencing and genotyping technologies, are expected to have an impact on breeding of specialty crops in the near future.

In many cases, research using one or another specialty crop is driven by important basic biological questions represented by that crop and the family of plants to which it belongs. As one example of many, the *Solanaceae* include closely related plants such as potato, tomato, eggplant, and many others domesticated for human use. That plant family is morphologically and physiological diverse and is thus a model to study the evolution of plant form and function. In addition, increasing consumer interest in locally produced and fresh market produce suggest that further investment in understanding their biology, physiology, diversity, and breeding is worthwhile.

Forest trees are America's largest renewable resource and are well poised to benefit from plant genomics research. With over 32 million acres, the United States leads the world in area of planted forests (AFPA 2007). The United States is the world's leading producer, consumer, and exporter of pulp and paper products. The annual value of wood used in manufacturing in 2002 was $20 to 30 billion (D. Adams, Oregon State University, personal communication, September 28, 2007). The industry is also reliant on renewable sources of energy; the pulp and paper industry supplies more then half of its energy needs, primarily through co-generation with its waste materials (DOE 2005a). Forest tree improvement is still a primitive discipline relative to breeding of the major crops. The issue is compounded by trees' long generation times, intolerance of inbreeding, high cost of vegetative propagation, large genome size, and recalcitrance to transformation. However, strong programs of cooperative tree breeding and genomics research are enabling the use of plant genomics information in forestry improvement, particularly for coniferous trees, such as the pines, and angiosperm species, such as poplar.

A central challenge for plant genome science is to understand how plants work at different levels of organization. This requires understanding how cells and organs make up an individual plant, and how that individual plant functions as a member of a community with other plants and microorganisms, all considered in ecological contexts that range from crop monocultures to highly diverse rain forests, across wide climatic ranges. Plant genome scientists seek to describe how plants grow from a fertilized egg and a small group of cells to a whole plant, shrub, or tree, as well as how they convert CO_2 and light into sugar and its derived products of economic value, including starch, protein, fiber, oil, and wood. Scientists also seek to define the evolutionary mechanisms by which species have adapted to the vast diversity of natural environments, producing an extraordinary diversity of forms. Their studies include how genetic variation within species allows them to mate, disperse, occupy wide geographic ranges, and persist over tens to hundreds of millions of years. The comparative power of plant genomics—whereby most of the genes, and the pathways they act in, can be rapidly compared between nearly any species—enables evolutionary lessons from model or wild species to directly inform plant improvement or genetic conservation efforts in crop species.

Plant genome sciences and enabling technologies are in a state of rapid development, leading to many new discoveries and applications. The precision, accuracy, and speed of basic plant genome technologies (such as basic DNA sequencing and gene expression analysis methods, and computational tools for analyzing and interpreting genomic data) continue to increase, and the costs of generating these data are dropping. These factors enable progress in the study of physiological mechanisms and organismal communities that were not previously tractable. The intensive use of plant models has led to dramatic progress in understanding of basic plant genomic biology. Twenty years after the adoption of *Arabidopsis* as a unifying model for plant biology—and the consequent birth of modern plant genomics research as embodied in the National Plant Genome Initiative (NPGI) and the National Science Foundation's (NSF) independent *Arabidopsis* 2010 Project—plant scientists are leveraging genomics technologies to accelerate the pace of basic discovery. In turn, the application of plant genome sciences to important societal problems has begun, though application has been slowed by social controversies in cases where genetic transformation was the avenue for translation.

THE NATIONAL PLANT GENOME INITIATIVE

History

The Interagency Working Group on Plant Genomes (IWG) was established in May 1997 by the Office of Science and Technology Policy (OSTP) under the direction of the National Science and Technology Council's Committee on Science, in recognition of the unprecedented scientific opportunities that plant genome research offered at that time. That recognition was predicated on the observation that very little crop genomics was being pursued in the public sector, and that crop genomics was being performed almost exclusively in a small number of corporations. Subsequent discussions between leading plant scientists, organized by the IWG, yielded a strong consensus that genomic tool development should be the highest priority for the early stages of NPGI, with a focus on the transitioning of model plant discoveries into crop species. The importance of basic research was driven home by the rapid success that had come from the choice of *Arabidopsis* as a model plant for study of the basic principles of plant growth and development. With *Arabidopsis* as a guide, it was clear that crop genomics could move quickly to real world applications.

IWG members, which included representatives from NSF, U.S. Department of Agriculture (USDA), U.S. Department of Energy (DOE), National Institutes of Health (NIH), Office of Management and Budget (OMB), and OSTP, were charged to: "(1) identify science-based priorities for a plant genome initiative; and (2) determine the best strategy for a coordinated Federal approach to supporting such an initiative, based on respective agency missions and capabilities." Subsequently, IWG developed a plan for a national plant genome initiative. The plan was approved by NSTC, and NPGI was officially established in 1998 as a coordinated national plant genome research project.

In the nine-year history of NPGI, members of the IWG have worked together to coordinate all activities in plant genome research among agencies to leverage resources and expertise. Since 1998, the membership of IWG has grown and currently includes the U.S. Agency for International Development (USAID) and the U.S. Forest Service (USFS), in addition to NSF, USDA, DOE, NIH, OMB, and OSTP. Plant genome research activities that existed prior to the inception of NPGI contributed to or have become part of NPGI. For example, NSF, USDA, and DOE were engaged in a $4-million-a-year project to sequence the genome of *Arabidopsis* before the inception of NPGI and this was incorporated into the goals of NPGI in 1998. The USDA Agricultural Research Service had maintained the maize stock center at the University of Illinois for many years, which has since expanded its operation to accommodate the growth of stocks resulting from NPGI activities.

During the 9 years of NGPI, and the nearly 20 years of focused and parallel investments in *Arabidopsis* research, U.S. plant science research has led the way internationally by any measure of productivity (see Appendix B). The total U.S. investment in competitive, peer-reviewed plant biology, including the flagship research programs, is less than $1 billion per year, roughly 30-fold less than comparable totals from NIH for programs focused on human health (Somerville 2006). Furthermore, U.S. leadership has, to date, often driven parallel investments by the

European Union, Japan, China, Australia, and other entities worldwide, creating a unique and academically rewarding international collaborative environment. In part because of this success, it is safe to say that without substantial growth from current funding levels, plant science research in the United States will soon trail that in Asia and Europe, likely leading to concomitant loss of competitiveness for U.S. science, technology, and plant agriculture.

Goals of NPGI

Initial Goals in 1998

The initial goal of NPGI was to understand the structure and function of every gene in plants with a focus on the species that are important to agriculture, environment, energy, and health. As stated in the 1998 NPGI plan, "This increased emphasis on the plant genome will radically change fundamental plant science research and its application to agriculture, forestry, energy, and the environment, as well as to the production of pharmaceuticals and other plant-based industrial chemicals and materials" (NSTC 1998). The scientific objectives for reaching those goals can be divided into three components:

- Genome structure—studies of the organization of genomes.
- Functional genomics—studies that relate genome structure and organization to plant function at the cellular, organismal, or evolutionary level.
- Application of the genomic information and knowledge for development of improved plants and novel plant-based products for human uses.

The initial plan was to invest in the first two components, thereby to provide linkages to the third component. The plan recommended increased federal investment to

- Accelerate the completion of the genome sequence of the model plant species *Arabidopsis thaliana.*
- Participate in an international effort to sequence rice.
- Develop the biological tools to study complex plant genomes such as corn, wheat, soybean, and cotton.
- Increase the knowledge of gene structure and function of important plant processes.
- Develop appropriate capabilities for handling and analyzing data.
- Ensure the accessibility of new information to the broader community of plant biologists.
- Maximize training opportunities that would arise from NPGI.

Goals in the 2003–2008 Plan

The objectives for 2003–2008 were to build on the scientific and technical advances of the first five years to ensure continued advancement in plant genomics and plant sciences. Technology development, data management and accessibility, and training of new generations of scientists were recognized as important goals. The six objectives were (NSTC 2003)

- Continued elucidation of genome structure and organization.
- Functional genomics—understanding the biological role of genomic sequences.
- Translational plant genomics—applications of genomic tools.
- Bioinformatics in every plant scientists' research toolbox.
- Education, training, and outreach.
- Consideration of broader impacts.

STUDY CHARGE AND SCOPE

NPGI will celebrate its 10th anniversary in 2008, and it is appropriate to assess what the initiative has achieved and to set goals for the future. IWG commissioned the National Research Council to convene a committee to assess the achievements of NPGI and recommend future research directions. The committee was charged to address the following:

- Review the accomplishments of NPGI to date.
- Assess the contribution of NPGI to science, research infrastructure, education of the next generation scientists, and international research collaboration.
- Discuss the broad impacts of NPGI to fundamental advances in biological sciences.
- Assess the contributions of NPGI to the application of scientific knowledge including technological innovation and economic competitiveness.
- Recommend future research directions and objectives for NPGI.

The committee was not to make budgetary recommendations.

Because the committee was charged to assess the contributions of NPGI to science, research infrastructure, education of the next generation of scientists, and international research collaboration, the committee conducted an in-depth assessment of research projects funded directly by NPGI participating agencies. The committee's assessment presented in Chapter 2 used three mechanisms. First, the committee analyzed the data provided by IWG on NPGI-funded projects. Second, the committee sent a questionnaire to all leading principal investigators of NPGI-funded projects (n=277) to ask them about the career paths of their trainees; their self-described most important contributions, published or otherwise, from their NPGI projects; their interactions with industry; and a list of all their publications citing NPGI funding (see Appendix C for the questionnaire). Third, the committee hosted a public workshop and invited plant scientists from universities, government agencies, and industry to solicit their evaluation of the achievements of NPGI in the last nine years and to discuss possible future directions of the program (see Appendix D for workshop agenda).

The scope of the study, however, was not limited to assessing only the funded research. The IWG has not only supported many activities and programs related to plant genomics over the last nine years, but also it has coordinated plant research among agencies. Research results from NPGI have been used to formulate mission-focused programs in participating agencies, as detailed in Chapter 2. Some IWG agencies also have provided, and continue to provide, databases, genomic technologies, sequencing facilities, and other in-kind support for plant genome research that IWG considers part of NPGI. A direct assessment of those contributions is difficult because there is not a clear definition of what endeavors are directly related to NPGI and which are ongoing within each IWG member agency that only peripherally support NPGI goals.

The committee attempted to take those NPGI-related activities into consideration, assessed whether NPGI has been achieving its goals, and recommends here future directions required to increase the impact of plant genome science research in the United States and around the world. In Chapter 3, the committee makes nine recommendations for NPGI on the basis of the contemporary societal issues facing the nation at present, the progress that NPGI has made to date and the areas that could be improved, and how NPGI could best achieve its goals.

2

Assessment

WHAT IS PLANT BIOLOGY RESEARCH IN 2007?

The ultimate goal of plant biology research and of the National Plant Genome Initiative (NPGI) is to create the knowledge-based capability to breed or produce plants with specific performance characteristics (phenotypic traits). Most traits of economic interest are under strong to moderate genetic control and are variable across populations and environments both within a species and between species. Discovering the genetic processes that control trait expression requires deep experimental knowledge in a few model species, intersected with broad knowledge of how natural variation in crop species and their close relatives contributes to it. Of course, the assumption that the most closely related genes across species share function is not always true, but it is an excellent starting assumption that is typically testable. Plant biologists aim to understand the "genetic wiring" of plants and of plant processes of basic, societal, or environmental interest. They aim to inform the breeding of plants with a defined genetic makeup, and to be able to predict with high certainty how these plants will perform, in different environments and climate conditions.

Examples of the traits that plant genome scientists would like to understand and control include disease resistance against a wide range of plant pathogens, nematodes, and insects and tolerance to environmental stresses (for example, salt, toxic soil chemistries, drought, extreme temperatures, and soil nutrient utilization). Other important targets include modulation of plant growth and development (for example, useful alterations of plant size, shape, and chemistry and the ability to use less fertilizer) and improved control of flowering and of the amount and quality of fruits and seeds produced (see Chapter 1).

Achieving the goals of breeding or producing plants with specific performance characteristics requires extensive investment in data generation, data management, and analysis infrastructures, and human capacity-building to make effective use of the data. It also requires a daunting level of intellectual growth in biologists' perception of how genetic networks control physiological traits, how natural genetic variability in important traits within and across plant species is manifested, how environmental signals are transduced into adaptive responses, and how evolutionary processes lead to network diversification, optimization, and creation of trait novelties.

SCIENTIFIC AND SOCIETAL IMPACTS OF NPGI

Impacts and Outcomes from NPGI-Funded Research

At the beginning of NPGI in 1998, there was little dedicated federal funding for plant genomics research beyond the then rapidly expanding *Arabidopsis* genome project and its associated research community, and various projects funded by ad hoc grants to principal investigators (PIs) from different research agencies. One exception was the U.S. Department of Agriculture's (USDA) National Research Initiative, which awarded 86 grants in FY 1997 worth about a total of $11 million from its "Plant Genomics" grant panel. These ad hoc efforts were split among many plant species, which arguably inhibited deep strategic investment in plant biology as a whole and genomics-based crop improvement in particular.

A fair assessment of NPGI, then, would address whether and how it has contributed to the building of strong and vibrant research communities linked by common interests. If these research communities have indeed been built, have they invested in cutting-edge genomic technology, and have they performed well using those resources? The committee relied on three key documents that articulated the goals (NRC 2002; NSTC 1998, 2003) and on the advice, critiques, and summaries of discussions at a workshop featuring key academic and private sector plant genome scientists (see Appendix D for workshop agenda and speakers). The committee also used data collected from a questionnaire sent to all lead principal investigators and reviewed the yearly NPGI Progress Reports (NSTC 1998, 1999, 2000, 2001, 2003, 2004, 2005, 2006, 2007).

The 5- and 10-year goals of NPGI were noted in Chapter 1 (see also NRC 2002; NSTC 1998, 2003). Some highlights of the research aimed toward those goals are emphasized in the following sections.

Capacity and Infrastructure Building

The committee views at least a significant part of the first nine years of NPGI as a capacity-building exercise, as also emphasized by the previous NRC report *The National Plant Genome Initiative: Objectives for 2003–2008* (NRC 2002). The capacity-building exercise was not trivial for two important reasons. First, there are many plant species, each of which might provide unique biology of interest to society. Hence, the mission of "plant genomics" is much broader than the mission of "animal genomics," which is nearly all driven by ultimate concerns for human health, and to a far lesser degree, by humans' uses of domesticated animals. Second, traditional plant biology research on the broad number of crops species took place in many institutions that, before NPGI began, had little exposure to either the mind frame or toolkit of genomics. The committee addressed how NPGI has built human capacity and how it has contributed to the distribution of a broad technological platform serving a variety of institutions and plant species. NPGI has done very well by those metrics.

First, the number of different PIs funded by NPGI grew nearly 13-fold over the first 9 years (from 21 to 277; see Table E-1 in Appendix E). As is perhaps expected, many of these PIs had more than one grant funded in that period. In sum, these numbers suggest that a critical mass of plant genomics PIs is being recruited for future efforts.

Second, the committee noted what seems at first glance to be a rather low proportion of investment ($14 million, or about 2 percent of the total) in the emerging, and often expensive, instruments required to compete effectively in genomics research (Table E-2 in Appendix E). The low investment in genomics instruments is partly a result of NPGI projects taking advantage of "sequencing for hire." Because sequencing for hire has become a lot cheaper over the nine-year course of the program, it results in cost savings compared to investing in large-scale sequencing equipment. Nevertheless, NPGI needs to ensure that its projects have access to the ever-changing landscape that characterizes high-throughput biology. Technology access facilitates previously impossible experimentation and in fact drives creation of new technologies. The rationale for further investment in technology access and technology creation in the framework of NPGI is discussed in detail in Chapter 3. Human capacity-building is addressed in the Education section below.

Genome Sequence, Structure, and Organization

NPGI has contributed to revolutionary breakthroughs in plant genome sequencing. The initial priority in plant genomics research is to have a high-quality finished genome sequence of the relevant organisms. The first such sequence for any species is referred to as the "reference" sequence (see below). NPGI initially invested in the international sequencing consortium that accelerated finishing of the *Arabidopsis thaliana* reference sequence (The Arabidopsis Genome Initiative 2000), and then helped to build an international consortium for sequencing the rice genome (see below). **The publications describing their genomes are citation classics.**

Because pathogens and pests cause great losses in crop yield, the sequencing of plant pathogenic genomes was included within the broader NPGI. Sequenced pathogens initially included bacterial pathogens (three strains of *Pseudomonas syringae*, three *Xanthomonads*, and several *Xylella* strains) and the fungal causative agent of rice blast, *Magneporthe grisea*. The NPGI subsequently supported the sequencing of additional fungal genomes, such as *Phytophthora* (three species that cause late blight of potato, root and stem rot of soybean, and sudden oak death syndrome; *http://www.oomycete.org/*). Three *Fusarium* species, three strains of *Verticillium* wilt, several powdery mildew and rust fungi, and the necrotrophic fungus *Botrytis cinerea* (Broad Institute 2007) were sequenced as part of a focus on fungi by the National Human Genome Research Institute (NHGRI). Genome sequences from additional pathogens, like *Hyaloperonospora parasitica* (oomycete causing downy mildew of *Arabidopsis*), are nearly finished. This first wave of plant pathogen genome sequences begins to cover the most economically critical plant pathogens, and it opens the door for comparative studies both across different isolates of one species and between species in the search for common mechanisms of virulence.

In addition to using "sequencing for hire" in some projects, NPGI has recently benefited from an extremely successful interaction with the Department of Energy's (DOE) Joint Genome Institute (JGI) to accelerate high-throughput plant and pathogen genome sequencing. That in-kind support to NPGI relies on a stringent external peer review by JGI that prioritizes projects on the basis of a mix of criteria, which include relevance to the DOE mission, organization and activity of the research community centered around candidate species, and evolutionary criteria aimed at maximizing the phylogenetic breadth of sampling. **It is the committee's view that the successful interaction of IWG with JGI, as the key (in fact, the only) major plant genome sequencing center, is critical to future overall success of NPGI.**

Comparative genomics is central to modern genetic approaches. Perhaps the most profound lesson of the Human Genome Project is that comparative analysis between closely and distantly related genomes provides a rapid and cost-effective way to extract information that can accelerate applied biomedical research and development. After the sequencing and analysis of the mouse and rat genomes (both model systems of direct relevance to biomedical research), it became evident that more sampling of diverse mammals would accelerate the identification and characterization of functional elements in the human genome through comparative analysis. That rationale led the NHGRI to sequence not only model organisms like the chicken and dog, but also the opossum, platypus, elephant, armadillo, and squirrel genomes. Nearly 20 mammalian reference genomes are either complete or in progress, totaling perhaps 60 billion base pairs. This rich comparative sequence landscape can lead to profound understanding of genome organization.

The development of plant comparative genomics has an important additional strength relative to the parallel comparative study of species related to humans. Humans, in essence, are the sole focus of biomedical research and this consideration drives the selection of relevant genomes to sequence. In contrast, there are dozens of societally important crops and wild plant species that are far more distantly related from one another than are mammals to each other. In addition, many species of plants have already undergone hundreds to thousands of years of domestication and agronomic improvement. They provide snapshots of how traits important to humans can be modified by selection. The multiple species focus of agriculture, therefore, places a premium on research approaches that can leverage generically useful genomics information for application across plant taxa. **The committee is confident that comparative genomics within and between plant families will accelerate the definition of gene function in parallel to the way comparative mammalian genomics has accelerated human genomics in the last five years.**

The evolutionarily conservation across plant genomes strengthens inferences made by comparative genomics methods. Hence, genome comparisons will have many useful cross-family applications between legume, rosaceous, solanaceous, and cereal crops, as well as between wood and fiber crops in the Salicaceae (poplar, willow), Myrtaceae (eucalypts), and the diverse families of conifers (pines and spruces). In particular, synergistic use of *Arabidopsis* and rice genome sequences can often allow definition of candidates for conserved function in other species for which, for example, expressed tag sequences (ESTs) from specific organs exist. There is also substantial and useful genome conservation that extends to the evolutionarily ancient gymnosperms, which include pine and spruce. Full-length cDNA clone sequences are even more useful to understand gene function and evolution; complete collections of full-length cDNA clones are important tools for subsequent functional experimentation.

Completed and Ongoing Land Plant Reference Genome Projects

Table 2-1 includes known ongoing plant genome sequencing projects, many of which support the mission of NPGI through in-kind support from JGI. The table includes only those projects that are expected to release sequences publicly in the

TABLE 2-1 Reference Plant Genomes Sequenced and in Progress

	Species	Common Name	Size (Gb)	Strategy	Estimated or Actual Date of Completion[a]	Group
1	*Arabidopsis thaliana*	Thale cress	0.2	BAC	2000	International consortium
2	*Oryza sativa* (x2)	Rice (*indica* and *japonica*)	0.4	BAC	2005	International consortium
3	*Populus trichocarpa*	Black cottonwood	0.5	WGS	2005	Joint Genome Institute
4	*Vitis vinifera*	Grape	0.5	WGS	2007	Genoscope
5	*Physcomitrella patens*	Club moss		WGS	2006	Joint Genome Institute
6	*Medicago truncatula*	Barrel medic	0.5	BAC[b]	2007	International consortium
7	*Sorghum bicolor*	Sorghum	0.7	WGS	2007	Joint Genome Institute
8	*Carica papaya*	Papaya	0.4	WGS	2007	University of Hawaii
9	*Ricinus communis*	Castor bean	0.4	WGS	2007	The Institute for Genomic Research
10	*Zea mays* (x2)	Maize	2.3	BAC[c]	2008	Washington University Genome Center
11	*Arabidopsis lyrata*	Rockcress	0.2	WGS	2007	Joint Genome Institute
12	*Selaginella mollendoerfii*	Spike moss	0.2	WGS	2008	Joint Genome Institute
13	*Mimulus guttatus*	Monkeyflower	0.5	WGS	2008	Joint Genome Institute
14	*Glycine max*	Soybean	1.1	WGS	2009	Joint Genome Institute
15	*Brachypodium distachyon*	Purple false brome	0.4	WGS	2008	Joint Genome Institute
16	*Prunus persica*	Peach	0.3	WGS	2008	Joint Genome Institute
17	*Solanum lycopersicum*	Tomato	1.0	BAC[b]	2010?	International consortium
18	*Brassica rapa*	Chinese cabbage	0.5	BAC	2009?	International consortium
19	*Capsella rubella*	Shepherds purse	0.2	WGS	2008	Joint Genome Institute
20	*Setaria italica*	Foxtail millet	0.5	WGS	2009	Joint Genome Institute
21	*Aquilegia formosa*	Western columbine	0.4	WGS	2009	Joint Genome Institute
22	*Eucalyptus grandis*	Eucalyptus	0.6	WGS	2009	Joint Genome Institute
23	*Lotus japonicus*	Trefoil	0.5	BAC	2010?	Kasuza DNA Research Institute

NOTE: The strategies used could be map-based sequencing using bacterial artificial chromosomes (BAC) or whole-genome shotgun sequencing (WGS).

[a]Several timelines in Table 2-1 are estimated from project websites or personal communication, and are hence approximate.

[b]BAC indicates only euchromatic BACs will be sequenced.

[c]BAC in addition to the BAC-by-BAC maize project, a second maize inbred line is being sequenced using a whole genome shotgun method by the Joint Genome Institute.

near future. A reference genome might have gaps and errors but captures greater than 90 to 95 percent of protein-coding gene content in highly accurate sequence (less than 1 error in 10,000 nucleotides), typically (but not always) anchored to physical and genetic maps. In some cases, targeted gap closure generates higher quality "finished" sequence. Resequencing projects that are aimed at characterizing variation relative to a reference sequence within a particular species, or efforts at finishing a nearly complete genome, are not included in this table. Some groups are still seeking funds to complete an ongoing sequencing project. Genome sizes are estimates of haploid content, given in billions of base pairs (Gb).

As noted above, the ultimate success of plant genomics is enriched by the knowledge of what genes are expressed in various cell types and organs under different stress conditions and their overall developmental time. The sampling of ESTs can give rise to a measure of the gene number (termed Unigene or, as in Table 2-2, a TIGR contig), and hence the deduced number of proteins, in an organism. Additional methods can sample the expression of the genome in specific tissues and cell types over developmental and environmentally altered conditions (transcriptomics). NPGI has contributed significantly to the collection of ESTs from various species, as shown in Table 2-2, and to the deployment of various transcriptomic tools.

Despite the large numbers of EST sequences and the equally compelling numbers of different cDNAs represented by these ESTs for many species, the extent and functional relevance of splicing of primary RNA transcripts and other elements and of alternate transcriptional starts and stops in plants are largely unknown. Whole genome analysis with tiling arrays using the *Arabidopsis* or rice genome sequences have made careful analysis in those important areas possible. Another calculation of the number of putative unique transcripts (PUTs) for these and other species can be found at the Plant Genome Database (Plant Genome Database 2007).

Gene Function, Expression, and Regulatory Networks

Genome sequence is the raw material for biological discovery. However, it is only one of the first steps toward understanding gene function, even at the biochemical level. In fact, plant scientists claim to have functional knowledge of only about 40 percent of the genes in *Arabidopsis*, and that estimate is based on an arguably overestimate of gene ontology (GO) functional inference. Hence, one important metric of plant genomics progress is whether the genomics tools have been generated with which to perform functional analysis in both high-throughput "data factories" and by hypothesis-driven studies of detailed gene function, usually in the laboratories of single investigators who specialize in functional networks of genes that act in a particular process or who study specific classes of genes.

NPGI has supported a wide range of investigations into gene regulatory mechanisms in model and crop plants. By virtue of the number of plant species funded by NPGI grants, the diversity represented by the funded projects is high. However, they generally fall into one or more of the following categories: defining

TABLE 2-2 Public Land Plant EST and Assembled Unigene of the National Center for Biotechnology Information (NCBI) or Contig Sequences of the Institute for Genomic Research (TIGR) That Are Deposited in Genbank up to August 2007

	Common Name	Total ESTs	Unigenes (NCBI)	TIGR Plant Transcript Assemblies
Eurosids II (Brassicas, citrus, cotton)				
Arabidopsis thaliana	Thale cress	1,276,692	29,918	27,983
Brassica napus	Oilseed rape	567,177	26,285	16,608
Gossypium hirsutum	Upland cotton	177,182	16,367	24,797
Citrus sinensis	Sweet orange	94,738	9,667	11,061
Gossypium raimondii	New world cotton	63,577	3,279	8,665
Citrus clementina	Clementine orange	62,250	6,106	5,222
Gossypium arboretum	Tree cotton	39,232	NA	4,591
Brassica rapa	Field mustard	33,316	NA	4,409
Brassica oleracea var. alboglabra	Wild cabbage	30,759	NA	6,761
Poncirus trifoliate	Japanese hardy orange	28,737	NA	5,083
Brassica oleracea	Wild cabbage	26,692	NA	See var. alboglabra above
Brassica rapa subsp. Pekinensis	Chinese cabbage	20,073	NA	4,409
Eurosids I (legumes, rosaceous plants, euphorbs, willows)				
Glycine max	Soybean	392,321	24,018	36,399
Malus x domestica	Apple tree	255,097	16,903	26,757
Medicago truncatula	Barrel medic	236,819	16,211	20,414
Lotus japonicus	Trefoil	150,631	13,640	14,461
Populus trichocarpa	Black cottonwood	89,943	14,059	12,687
Populus tremula x Populus tremuloides	Hybrid aspen	76,160	7,519	11,593
Prunus persica	Peach	70,972	6,306	6,596
Ricinus communis	Castor bean	53,402	NA	4,524
Populus trichocarpa x *Populus deltoides*	Hybrid poplar	53,208	NA	7,803
Euphorbia esula	Leafy spurge	47,543	NA	9,905
Arachis hypogaea	Peanut	40,627	NA	1,491
Trifolium pratense	Rotklee clover	38,109	NA	4,347
Populus tremula	European aspen	37,313	NA	5,961
Manihot esculenta	Cassava	36,120	NA	5,189
Phaseolus vulgaris	Common bean	22,847	NA	2,941
Bruguiera gymnorrhiza	Burma mangrove	20,373	NA	2,031
Populus trichocarpa x *Populus nigra*	Hybrid poplar	20,130	NA	3,531
Phaseolus coccineus	Scarlet runner bean	20,120	NA	2,315

continued

TABLE 2-2 Continued

	Common Name	Total ESTs	Unigenes (NCBI)	TIGR Plant Transcript Assemblies
Asterids				
Solanum lycopersicum	Tomato	257,093	16,945	21,523
Solanum tuberosum	Potato	227,289	19,539	26,280
Helianthus annuus	Common sunflower	94,111	7,955	10,219
Nicotiana tabacum	Common tobacco	88,579	8,436	10,693
Lactuca sativa	Garden lettuce	80,781	7,839	11,215
Ipomoea nil	Japanese morning glory	62,282	NA	11,216
Coffea canephora	Robusta coffee	55,692	NA	6,732
Lactuca serriola	Prickly lettuce	55,490	NA	7,125
Cichorium intybus	Chicory	41,747	NA	6,501
Nicotiana benthamiana	Tobacco	41,440	NA	4,836
Taraxacum officinale	Dandelion	41,296	NA	5,993
Helianthus tuberosus	Jerusalem artichoke	40,362	NA	5,845
Helianthus exilis	Serpentine sunflower	33,961	NA	5,187
Capsicum annuum	Pepper	31,090	NA	4,189
Lactuca saligna	Willowleaf lettuce	30,696	NA	4,999
Helianthus paradoxus	Paradox sunflower	30,517	NA	3,864
Cichorium endivia	Endive	30,171	NA	4,098
Lactuca virosa	Wild lettuce	30,068	NA	4,912
Lactuca perennis	Wild lettuce	29,125	NA	4,485
Helianthus petiolaris	Prairie sunflower	27,484	NA	3,994
Antirrhinum majus	Snapdragon	25,310	NA	4,221
Ocimum basilicum	Sweet basil	23,260	NA	3,343
Helianthus ciliaris	Texas blueweed	21,590	NA	3,070
Other eudicots				
Vitis vinifera	Grape	320,538	22,278	21,627
Aquilegia formosa x Aquilegia pubescens	Western columbine	85,039	7,555	12,160
Mesembryanthemum crystallinum	Common ice plant, "basal" core eudicot	27,348	NA	2,897
Beta vulgaris	Beet, "basal" core eudicot	26,745	NA	3,868
Monocots (includes grasses)				
Oryza sativa	Rice	1,211,418	40,259	49,870
Zea mays	Maize	1,159,264	57,447	64,601
Triticum aestivum	Wheat	1,050,926	34,505	62,121
Hordeum vulgare + subsp. vulgare	Barley	437,713	21,418	30,171
Saccharum officinarum	Sugarcane	246,301	15,586	26,894
Sorghum bicolor	Sorghum	204,308	13,547	20,714
Festuca arundinacea	Fescue	41,869	NA	6,297
Zingiber officinale	Ginger	38,139	NA	7,850
Hordeum vulgare subsp. spontaneum	Barley	24,161	NA	See sp. vulgare above

TABLE 2-2 Continued

	Common Name	Total ESTs	Unigenes (NCBI)	TIGR Plant Transcript Assemblies
Sorghum propinquum	Sorghum	20,881	NA	3,402
Brachypodium distachyon	Purple false brome	20,449	NA	2,785
Allium cepa	Onion	20,159	NA	3,578
Gymnosperms				
Pinus taeda	Loblolly pine	328,628	18,859	28,060
Picea sitchensis	Sitka spruce	139,569	15,683	11,551
Picea glauca	White spruce	132,623	17,810	16,102
Picea engelmannii x Picea glauca	Hybrid spruce	28,170	NA	5,767
Pinus pinaster	Maritime pine	27,288	NA	3,901
Other land plants				
Physcomitrella patens subsp. Patens	Moss	174,908	13,688	18,707
Marchantia polymorpha	Liverwort	33,692	NA	3,874

NOTE: All plants with more than 20,000 ESTs are shown, as listed in dbEST.

gene function, defining regulatory genes and networks, understanding patterns of gene expression, comparative analysis of gene expression, gene expression resources and databases, and epigenetics and RNA-based regulation. This information is captured in both the published record and in various databases (see Appendix F). A brief summary of the many highlights includes the following:

Defining gene function. Large-scale insertional mutagenesis and TILLING resources, first deployed in *Arabidopsis* but now available in a variety of crop species, have revealed key functional and phenotypic knowledge and provided vital resources for further work (see Table 2-3). The ability to define gene function via loss of function mutation remains the bedrock of genomics, and methods to overcome genetic redundancy and other impeding factors are further being developed. Those methods include the engineering of artificial micro-RNAs capable of silencing several members of a gene family simultaneously.

Defining regulatory genes and networks. Several projects focused on identification of novel regulatory genes and features through genome-wide approaches. For example, regulatory networks and factors that control host-microbe interactions and disease resistance, largely identified by large-scale forward genetics in *Arabidopsis*, are now being exploited in rice, tomato, and legumes, among others, using forward and reverse genetics methods enabled by genome sequences.

Understanding patterns of gene expression. Functional genomic technologies were developed and applied to analyze gene expression patterns in different cell types, tissues, and organs, and in plants under stress and undergoing developmental transitions. Microarray and other high-throughput profiling tools have been used to identify and characterize important genes for root architecture, leaf form, and tomato fruit development, just to name a few examples. Many of these projects have yielded publicly available expression atlases and searchable resources.

Comparative analysis of gene expression. Comparative analysis of expression patterns could be a major outcome of functional genomics applied to a wide variety of plant species. Some success has been realized in NPGI-funded analysis of genes involved in flower development across an evolutionary spectrum of plants. Genes that are regulated by the circadian clock, and by photoperiodic regulatory modules, are being revealed through comparative profiling and analysis in *Arabidopsis*, poplar, and rice.

Gene expression resources and databases. Several databases and online resources emerged from NPGI-funded projects (Table 2-3). Those resources include the MPSS database of transcript and small RNA expression data, and the PlexDB database for expression data. Many of those resources are used regularly by PIs of NPGI projects. (See the list of Websites that NPGI PIs reported as their five most-used websites for their work in Appendix F.)

Epigenetics and RNA-based regulation. The diversity and functions of small RNAs (20–25 nt) that affect both genic and intergenic sequences have been revealed using innovative high-throughput sequencing technology in a variety of dicot and monocot models and crops. This work will enable a more subtle understanding of gene regulation and the evolution of developmental regulator processes. NPGI-funded projects have contributed to the rapidly expanding field of epigenetics, which deals with heritable changes and patterns that occur without changes in DNA sequence. Epigenetic inheritance properties are controlled by the structure of chromatin as expressed in changes to histones and DNA methylation, which are affected by polyploidy, hybridization, and the expression of small RNAs. Genome-wide surveys and functional analysis of genes affecting epigenetic inheritances have been done in maize, *Arabidopsis*, and a few other species.

Shortcomings in Gene Function, Expression, and Regulatory Network Analyses

Not all progress that was envisioned five years ago (NRC 2002) has been realized. Integration of data across plant species remains a considerable challenge, partly because of the heterogeneity of datasets, disparate data standards, lack of sufficient experimental tools, and small number of groups funded to do database and experimental integration work. **Data integration across heterogeneous plat-**

TABLE 2-3 Examples of websites that are direct results of NPGI and IWG collaborations.

Title	Website	Estimated Number of Unique Visitors in a Month[a]	Estimated Number of Page Impressions in a Month[a]
The *Arabidopsis* information resource	*http://www.arabidopsis.org/*	33,795	939,272
The Barley Coordinated Agricultural Project	*http://barleycap.org/*	253	1,454
Dendrome: A Forest Tree Genomics Database	*http://dendrome.ucdavis.edu/*	3,600	30,000
The Prunus Genome Database: A Model for Rosaceae	*http://www.bioinfo.wsu.edu/gdr//*		
GrainGenes: A Database for Triticeae and Avena[b]	*http://wheat.pw.usda.gov/GG2/*	32,256	419,632
Gramene	*http://www.gramene.org/*	607,966	4,786,295
The Legume Information System	*http://www.comparative-legumes.org/*	4,628	66,997
Maize Genetics and Genomics Database	*http://www.maizegdb.org/*	30,358	179,671
The Maize Tilling Project	*http://genome.purdue.edu/maizetilling/*	452	2,196
The Floral Genome Project	*http://www.floralgenome.org/*		
The Cotton Genome Database	*http://www.cottondb.org/*	2,111	287,857
Comparative cDNA Sequencing in Radish (Raphanus)	*http://radish.plantbiology.msu.edu/*	65	175
The Populus Genome Portal	*http://genome.jgi-psf.org/Poptr1/Poptr1.home.html*		
PlexDB: A community resource for plant and plant-pathogen microarrays	*http://www.plexdb.org/*		
Rice Coordinated Agricultural Project	*http://www.uark.edu/ua/ricecap/*	4,044	40,365
Rice MPSS	*http://mpss.udel.edu/rice/*		
Soybase and the Soybean Breeders' Toolbox	*http://soybase.org* *http://soybeanbreederstoolbox.org/* *http://soybeanphysicalmap.org/*	2,104	24,413
Wheat Coordinated Agricultural Project	*http://maswheat.ucdavis.edu/*	566	1,555
Potato Functional Genomics	*http://www.potatogenome.org*	4,777	62,775

[a]The estimated number of unique visitors and page impressions do not include robot requests. They were provided by managers or principal investigators of the websites. Estimates were not available for all Websites because the statistics were not kept for some or because the web managers or principal investigator did not respond to the request.

[b]The GrainGenes database has a mirror Website (*http://grain.jouy.inra.fr/GG2/index.shtml*) that also is actively used. That website has an estimated 25,975 unique visitors and 269,323 page impressions, excluding robot requests.

SOURCE: Examples of Websites were obtained from the Interagency Working Group on Plant Genomes.

forms and metadata types presents a difficult challenge to efficient extraction of knowledge and translation to crop species.

Genetic redundancy, where multiple genes serve overlapping functions, has limited some functional analysis of genes in some high-throughput phenotype screens. The extent of genetic redundancy was not recognized in the early years of NPGI, and future projects need to take redundancy into account during experimental design. Further, it is now known that epigenetic regulation is masking considerable phenotype expression. Revealing the full genetic potential of plants, therefore, will require future studies that unmask hidden phenotypes under epigenetic control.

Finally, even for the most advanced plant models, technical limitations have yet to be overcome in a number of areas. For example, the technology to consistently and predictably perform gene replacement is not generally available. It is also not feasible to do large-scale capture of protein interactomes using large sets of epitope-tagged proteins. For nearly all species, the important tool of genetic transformation—namely the ability to generate large numbers (over thousands) of traditional transformants—is expensive and not well developed.

Domestication, Diversity, and Natural Genetic Variation

The 2002 NRC report noted that while sequencing costs were too high for deep sequencing across the plant kingdom, EST sampling, development of mapping tools, and the coalescence of focused, community-supported projects of evolutionary and ecological interest were worthy of support. NPGI-funded studies of natural variation and crop domestication have led the way in the understanding of selection and in the dissection of complex traits. Results from those studies have been influential in human genetics as well (see example below). Further, as noted above, JGI has been the focal point for communities of scientists to develop genomics-based programs to understand unique and important concepts in evolutionary and ecological genomics using *Mimulus* (Monkey flower) and *Aquilegia* (Columbine). Important scientific results from NPGI funding include:

- Initiation of "association mapping" (population-based mapping of traits inherited in a complex or multigenic manner) in plants. Association mapping facilitates identification of the actual genes that underlie Quantitative Trait Loci (QTL) mapping down to the gene level. The examples funded by NPGI are the first to explicitly use structured association mapping in any species.
- Development of leading statistical genetic algorithms for association mapping, which outperform many of those developed for human biomedical research, and have been adopted by the human genetics community.

• Provision of detailed examples of the genetic events that led to crop domestication—these include molecular details, the dynamics of natural selection, and the ultimate effects of those genetic changes on the domesticated trait.

• Identification of reservoirs of natural diversity and core germplasm in a variety of crops.

• Creation of key resources for mapping complex traits, including the maize nested association mapping panel, which when released at the end of 2007 will be the largest complex trait dissection system for any species.

• Acceleration of positional cloning of QTL for many species.

Informatics, Modeling, and the Virtual Plant

Clearly, locating the information generated by a program as large and diverse as NPGI would be an overwhelming, time-consuming task without interfaces that are easy enough to use routinely. As a result of NPGI support, many websites and databases were expanded or newly developed (listed in Table 2-3). An important positive note is that several of these receive very large numbers of visits. Further, a key organizing principle is that many of these sites are sharing software—something that is desirable but considered by many to be difficult or impossible. Software reuse has the dual benefits of reducing costs (the software is only developed once) while improving quality (because users and informatics support staff from all sites sharing the software can critique and debug). The Generic Model Organism Database (GMOD 2007) project, with funding contributed by NPGI, is one of the reasons the software re-use is possible. The aim of GMOD is to collaboratively design and build a database architecture and constituent software components for creating and managing genome-scale biological databases. A number of the projects listed below are active GMOD contributors or users in Table 2-3. Some share a common database schema and associated utilities, and others are using the same interface application (for example, cMAP).

An additional collaborative success is the direct, active involvement of GMOD contributors in developing semantic community standards. Through their efforts, shared anatomical, developmental, phenotypic, and other controlled vocabularies and ontologies are being developed and maintained. It is only with clear and accepted standardization that semantically driven mechanisms for data mining, integration, assimilation, knowledge discovery, and analysis become possible, and many of the projects in Table 2-3 are building this foundation.

The available plant genomics databases have already led to concrete benefits in the plant research community, as summarized above. From an analytical perspective, the gains are clear: Collection of large data sets from multiple sources has led to metadata analyses, though comparisons of data that span technology or plant physiological platforms have to be done with caution. The utility of the databases to formulate testable experimental hypotheses is most advanced in the plant systems where there are most data. Here, genome-wide based analyses can have predictive value, and therefore can be used to prioritize targets for future experimentation. For example, examining the joint probability of observing particular DNA regulatory sequence motifs (representative of binding sites for a particular transcription factor family) together with analysis of their expression under certain environmental stresses could lead to new insights of co-regulatory nodes. Grass genomes are genetically well aligned, and traits that are distinctive to each species can be selected for further studies. Comparative studies exploiting the basal plant genomes now available provide the basis for phylogenetic models of gene and genome structure. To enable these and other studies, the data resource projects typically offer downloading datasets such as physical maps, genetics maps, computed gene predictions, microsatellite sequences, the database content (as tab-separated-value files or as database dumps), protein or nucleotide sequences, and curated, versioned gene models.

The community has adopted these resources, and their use has dramatically increased over the years, as evidenced by the increase in database visits at the most heavily used Websites. Currently, the two limiting factors are the problems presented by disparate, heterogeneous datasets that make comparisons across databases difficult, and the lack of students and postdoctoral fellows trained in computational skills and statistics. Without a skilled workforce to use them, the value of database resources and large datasets cannot be fully realized. Efforts are being made by the databases that serve large communities (for example, TAIR and Gramene; see section on Education below) to educate users on access to the resources available at the database Website.

The availability of genome-based information has set the stage for improved modeling and understanding of biological processes from the cellular, to organismal, and ultimately to the level of entire plant communities. At present, in any given organism, the function for roughly 30 to 50 percent of the genes can be inferred, but these inferences are only guides for subsequent experimental proof of function. The function of the remainder is still unknown, and gaining that knowledge is a challenge for upcoming years. However, plant biologists can place the genomic knowledge gained into an integrated systems framework. Metabolic pathway information is available for both *Arabidopsis* (TAIR 2007) and rice (Gramene 2007). These are both implementations of bioCyc, which is another example of a GMOD component.

At the scale of the organismal level, sophisticated image analysis techniques of gene expression are being used to understand the roles of relevant genes and the environmental factors that influence the developing meristem in *Arabidopsis* (see Scientific Inference Systems Laboratory 2007). This project is pioneering techniques for constructing cellular models of coordinated patterns of gene expression and will enable simulation of developmental processes under different conditions.

Translational Genomics

The 2002 NRC report recommended an expansion of NPGI research into areas that enable the translation of findings from reference species to related crop plants—known as *Translational Agriculture.* That newly proposed area of research was intended to empower both public and private scientists to use the basic discoveries from other research areas of NPGI for crop improvement. The report mentioned the following as critical enabling technologies and goals: the development of genetic maps, physical maps, transcript maps, and germplasm collections with molecular genotypes; the discovery of chromosomal intervals that possess gene conditioning important agronomic, compositional, and pest resistance traits; the design of "breeder-friendly" DNA markers; and the analysis of fungal genomes and the sequencing of the gene-rich regions of major crop plants.

This committee found important progress in almost all those areas for some crops. For example, major advances in the small-grain crops of wheat, barley, and rice include:

- The nationally coordinated application of DNA markers in publicly funded wheat breeding programs for selection of quality and pest resistance traits (University of California Davis 2007).
- The positional cloning (relying on DNA markers, genetic maps, large-insert DNA libraries, and a physical map of the critical genomic region) of the genes underlying vernalization in wheat and barley (Fu et al. 2005; Yan et al. 2006; Yan et al. 2004; Yan et al. 2003).
- The mapping of *Gpc-B1*, a gene from a wild wheat accession that increased grain protein content in cultivated wheat (Distelfeld et al. 2006).
- The cloning of a gene for stem rust (a major disease of both wheat and barley) resistance from barley (Brueggeman et al. 2002).
- The provision of industry-wide training by rice breeders in the application of marker-assisted selection technologies (University of Arkansas 2007). In addition, the committee noted that NPGI discoveries have led to a variety of productive interactions with the private sector, including a handful of new, start-up companies. These are mentioned in Appendix J in the PIs' responses about "Collaborations with Industry."

DNA marker technologies have become widely employed in the improvement programs of major U.S. commodity crops. NPGI has enabled the development of more efficient DNA markers for that purpose. The NPGI funding has supported the development of BAC libraries for several crops and research to obtain BAC-end sequence to arrange overlapping BACs in their proper location within a contig. That sequence information has been used as a template for the discovery of genome-wide markers for single-nucleotide polymorphism (SNP). The identification of the SNP markers for major commodity crops, along with their adaptability to high-throughput platforms and reduced cost per data point, has led to their rapid introgression in crop improvement programs and provides a clear example of the leveraging and translational power of the NPGI resource development. NPGI also has had a major impact on how crops are bred in the private sector. During the last five years it has become common practice for commercial crop breeders to employ DNA-marker assisted selection, particularly in the development of improved corn inbreds and soybean cultivars.

NPGI has recognized the need for translational genomics applicable to trees. For example, poplar serves as a model organism for genomic studies of tree and wood development, and bioenergy feedstocks because of its modest genome size (about 480 Mb) and facile capacity for transformation (Brunner et al. 2004). The poplar genome was sequenced by JGI and an international consortium of collaborators (Tuskan et al. 2006). The highly outbred genetic structure of many forest trees make linkage disequilibrium blocks extremely small (Brown et al. 2004), increasing the accuracy of association genetic studies (Neale and Savolainen 2004). In order to understand the nature of natural variation in trees and to develop useful markers, more than 8,000 genes have already been resequenced and SNPs discovered in a major NPGI-funded study of loblolly pine wood properties. Earlier studies identified a null allele of the cinnamyl dehydrogenase gene (*cad-n1*) as the largest-effect major gene known for volume growth in loblolly pine (Yu and Buckler 2006). NPGI-funded translational projects oriented toward developing tools for ecological studies and restoration of disease-damaged wild trees include an effort to save the American chestnut, by breeding or engineering varieties resistant to the introduced chestnut blight pathogen. Other research funded by NPGI grants has led to genetic markers useful for DNA fingerprinting of clones to aid management during breeding and propagation, and have served as useful markers for ecological studies of gene flow and community ecology (for example, Whitham et al. 2006).

The NPGI Literature Footprint

NPGI and the independent National Science Foundation's (NSF) *Arabidopsis* 2010 Project are the engines of basic plant genomics discovery that power intellectual and practical furtherance of plant biology in both the public and private sectors. Hence, the committee also monitored scientific output of NPGI-funded grants (as provided by IWG) using the traditional metric of peer-reviewed publications and their impact on international plant science. The data in Appendix B were gathered based on surveying the 40 most cultivated crop species by area harvested (FAOSTAT 2007). The committee assessed whether the investment made by NPGI had increased the knowledge footprint for each species on the basis of these data. For each species, the Web of Science—a database of about 8,700 journals—was queried for articles with each crop's species name or common name in the title. To narrow down the list of articles to those relevant to genomics, only the articles that include either *genomics, genomic, sequencing,* or *sequence* as a keyword are included. These data are also compared to plants that serve as broad models for all of basic plant biology (bottom of the tables in Appendix B).

The numbers of laboratories contributing to publications using the model plants that were the focus of NPGI has, in most cases, grown considerably, as have the number of laboratories using most models, including *Arabidopsis.* Furthermore, the U.S. contribution to the top 10 most-cited papers for many species has remained high or increased. Finally, the percentage of total publications that included U.S. institutions has risen or stayed constant for all of the most important crop species except, notably, soybean.

The committee also measured the publication impact of NPGI over the nine-year period (see Appendix G). The 165 PIs who responded to the committee's questionnaire (from 277 total; shown in Appendix C) cited 1,478 peer-reviewed publications that are included in the 2006 ISI Journal Citation Impacts. The committee noted that 317 of these publications (or 21 percent of the total) appeared in highly cited journals across all of life sciences journals (citation impact of 9 or higher). This striking finding suggests that NPGI is generating novel, important, and topical science. A total of 659 publications (or 45 percent of the total) appeared in journals with citation impact of 6 or higher. **The committee concludes from these data that many NPGI PIs are producing highly competitive publications that are published in the most outstanding scientific journals; this is a major accomplishment for NPGI.**

Education

In the 2003–2008 NPGI plan (NSTC 2003), the roles of education, training, and outreach in realizing the full potential of NPGI for plant sciences within the United States were emphasized. Five goals were outlined:

- Traineeships for undergraduates, graduate students, and postdoctoral researchers in plant genomics research.
- Informatics training for both established and young investigators.
- Mid-career training programs in plant genomics for university and college faculty and plant science professionals.

- Workshops to inform the broader research community about accessing and using the NPGI research resources.
- Outreach to the K-12 community.

Some aspects of these objectives have been partially met, but others have yet to be fulfilled. The following sections discuss progress to date.

Undergraduate, Graduate, and Postdoctoral Traineeships

There has been substantial investment in undergraduate, graduate, and postdoctoral stipends funded through NPGI (Tables 2-4, 2-5, and 2-6). There are no

TABLE 2-4 Number of Undergraduate Students Trained and the Sector in Which They Now Hold Positions, as Reported by the 165 NPGI-Funded Investigators Who Responded to the Committee's Questionnaire

Sector	Number of Students
Academia, including undergraduate and graduate schools and other positions in academia	438
Government	8
Industry	79
Still in investigators' laboratory	198
Other, including professional schools such as medical, dental, and law school	115
Unknown	658
Total	1496

NOTE: Number of undergraduate students among the 1,496 known to have left the country = 10.

TABLE 2-5 Number of Graduate Students Trained and the Sector in Which They Now Hold Positions, as Reported by the 165 NPGI-Funded Investigators Who Responded to the Committee's Questionnaire

Sector	Number of Students
Academia, including undergraduate and graduate schools and other positions in academia	214
Government	25
Industry	60
Still in investigators' laboratory	190
Other, including professional schools such as medical, dental, and law school	25
Unknown	60
Total	574

NOTE: Number of graduate students among the 574 known to have left the country = 45.

TABLE 2-6 Number of Postdoctoral Researchers Trained and the Sector in Which They Now Hold Positions, as Reported by the 165 NPGI-Funded Investigators Who Responded to the Committee's Questionnaire

Sector	Number of Students
Academia	318
Government	46
Industry	58
Research in other sector (for example, nonprofit)	17
Still in investigators' laboratory	193
Other	21
Unknown	64
Total	717

NOTE: Number of postdoctoral researchers among the 717 known to have left the country = 129.

readily available data to indicate how many of these trainees are or were U.S. citizens, whether they stay in plant genomics after the completion of their training, or whether NPGI funding attracted previously uncommitted students into plant genomics. The lack of long-term follow-up data on the students' career paths compromises the committee's ability to assess the direct impact of NPGI education funding on the future of plant genomics. Nonetheless, NPGI is exposing an increasingly large number of students to the "excitement of scientific discovery in a field at the cutting edge of biology" (NSTC 2003).

NPGI projects have supported large numbers of students and postdoctoral researchers. However, no specific plant genomics graduate or postdoctoral fellowship programs or "cross-over" fellowship programs to bring non-plant scientists (particularly those with quantitative or computational skills) into plant genomics were created or funded. Specialized individual fellowship programs provide incentive for the best students at any level to seek and win competitive fellowships and have proven to bring new talent into a field. For example, the NSF Plant Biology Post-doctoral Fellowship Program (1983–1994), which was not a part of NPGI, recruited 237 scientists into plant research, nearly all of whom were trained in other fields of biology. This program, especially in the beginning, populated plant science with young researchers trained in the molecular biology of well-developed systems. Several members of this cohort are now members of the National Academy of Sciences; one was a Howard Hughes Medical Institute Fellow and is now the chief executive officer of a start-up company. Many more are productive faculty members or high-level administrators at institutions of all types, governmental science policy makers, and corporate scientists. Plant sciences have reached yet another turning point, where a similar infusion of scientists trained in quantitative disciplines like computer science, applied mathematics, statistics, biomedical engineering, quantitative genetics and plant breeding, and ecological or evolutionary biology would be welcome additions to most plant genome projects.

Since 2005, NSF has funded three new graduate programs relevant to NPGI through its Integrative Graduate Education and Research Traineeship (IGERT) program. In FY 2005, the University of California, Riverside Center for Plant Cell Biology and the University of California, San Diego received IGERT awards for doctoral training programs in chemical genomics and in plant systems biology, respectively; both programs will bring together faculty in plant cell biology, chemistry, computational biology, and engineering (NSTC 2006). A new IGERT program at the University of Arizona focuses on evolutionary, functional, and computational genomics (University of Arizona 2007). Although the numbers of students involved in those IGERT programs are low, the three programs are a first of what could be an expanding set of steps towards addressing a key recommendation of the 2002 NRC report that "the plant biology community needs to expand training opportunities into disciplines that are not traditionally associated with plant biology and crop sciences, such as computer science, mathematics, chemistry, and engineering."

At the undergraduate level, NPGI provides partial support for 12 summer internship programs through its Research Experience for Undergraduate Program (REU) (NSF 2006). In each program, about six to ten students chosen through a

BOX 2-1
Genome Consortium for Active Teaching

Partial funding from NPGI has supported the Genome Consortium for Active Teaching (GCAT) (Davidson College 2005) to bring functional genomics methods into undergraduate curricula by providing undergraduates with access to affordable microarray technology, including the arrays, scanning services, free software for data analysis, and faculty workshops. In the first seven years of the project, 5,000 microarrays provided by GCAT have been used by 141 faculty and 6,000 students on 134 campuses. By 2009, GCAT estimates that 9,480 undergraduates per year will have access to its microarrays (Campbell et al. 2007). NPGI supports best practices workshops on the use of the microarrays and software for faculty at primarily undergraduate and minority-serving institutions (NSTC 2006), and also enables GCAT to maintain a helpdesk, staffed by students, for users of the microarray data analysis software package developed by GCAT faculty and students (Campbell et al. 2007). It is too early to assess whether this program influenced subsequent career choices for these students. However, preterm and postterm surveys of student participants documented substantial gains in their knowledge about microarray experiment design, error, gene expression, clustering analyses, and interpreting microarray results. Surveys and open-ended responses from faculty members who attended the workshops also revealed positive impacts on their teaching programs (Campbell et al. 2007).

BOX 2-2
Reaching Out to Underrepresented Populations

Several NPGI-funded programs specifically target underrepresented populations of high school and college students. For example, Iowa State University and the USDA Agricultural Research Service (USDA-ARS) Station in Ames, Iowa, offer Native American students eight-week summer research internships that focus on genetic and bioinformatic investigations of diversity among plants with cultural and historical importance to Native Americans (NSTC 2007). The multi-institutional Research Experiences for Undergraduates (REU) programs led by North Carolina State University and by the University of Connecticut offer opportunities for diverse populations of students to gain hands-on experience in plant genomics research (NSTC 2005). An interdisciplinary team of scientists from the University of Wyoming, The Institute for Genomic Research (TIGR), and Cold Spring Harbor Laboratory provide enhanced educational and research opportunities through exchange visits, courses, and hands-on workshops for both faculty members and students from Little Big Horn College, a tribal community college in Crow Agency, Montana (NSTC 2007). Several NPGI-funded laboratories have entered into formal partnerships with historically black colleges and universities, in which African American masters' candidates and undergraduate students participate in the research at the host laboratories over the summer and take their projects back to their home institutions during the academic year (NSTC 2001). It is not clear whether these programs influenced subsequent career choices for these students; assessment methods need to be developed and deployed.

competitive national search receive a stipend and housing for eight to ten weeks. In addition, at least 14 faculty members from primarily undergraduate institutions were hosted by NPGI-funded laboratories through NSF's Research Opportunity Award program. Several opportunities also exist for high school students and science teachers to participate in NPGI-funded research through NSF Research Assistantships for High School Students or Research Experiences for Teachers supplements to individual laboratories. Boxes 2-1 and 2-2 provide additional examples of the NPGI contribution to undergraduate education.

Informatics and Mid-career Training

In contrast with the growing opportunities for graduate training and the slate of undergraduate educational programs, the rate of progress in providing training opportunities in bioinformatics for established and new investigators or in plant genomics for plant breeders and physiologists has been considerably slower. The 2002 NRC report proposed a national strategy for bioinformatics that included training, collaboration with large data centers, and bioinformatics-oriented research. Strategies to address this perceived gap were also presented as key objectives of the proposed Plant Cyberinfrastructure Center (Meyerowitz and Rhee 2006).

Although the anticipated new generation of researchers specializing in plant genomics is emerging, there is a need for experienced plant physiologists and plant breeders who have acquired skills in genomic technologies. The lack of plant breeders who are well versed in genomic approaches is seen as a major impediment to translational plant genomics and to the future of plant improvement in the public and private sector in the United States. NPGI-supported workshops on marker-assisted selection for plant breeders are a good start to correcting this deficit (NSTC 2006). Outreach to plant breeders, and potentially to farmers, seems to be within the mandate of USDA and its extension arm, but it is unclear whether there has been a concerted effort in this regard. In the first year of the Wheat Coordinated Agricultural Project (CAP), USDA provided workshops or information sessions on marker-assisted selection at more than 40 field days and industry meetings, and mounted a symposium at the Crop Science Society of America meeting that reached more than 120 people (USDA-CSREES presentation to the committee, April 26, 2007, Workshop). NPGI has also sponsored workshops at the Plant and Animal Genome Conferences on specialized topics relevant to specific crops and on general subjects such as database construction, transcriptional profiling, and genomic computing (NSTC 2000).

Informing the Broader Research Community

The 2002 NRC report called for organizers of community databases to improve the user skill level through short courses and exchange visits (NRC 2002). For example, the *Arabidopsis* Information Resource (TAIR) has offered one-hour to two-hour introductory and advanced workshops at Plant and Animal Genome Conferences, the International Conference on *Arabidopsis* Research, and the American Society of Plant Biologists Meeting. TAIR usage has increased steadily since the project was founded in 1999.

K-12 Education and Outreach

Some NPGI grantees have invested considerable energy in developing outreach efforts targeted towards K-12 students. Several held workshops, where K-12 teachers learn about genomics and biotechnology and develop their own curricular modules or lesson plans (see Appendix H). Recognizing that most precollege teachers are not trained in the practice of science as a process, at least two exemplary NPGI-funded programs provide six- to eight-week full-time mentored research internships through which teachers gain first-hand experience in a plant genomics laboratory, as well as education in current learning theory research, so that the teachers are well-equipped to develop research-based curricula. Other outreach efforts resulted in Internet-accessible activities and kits (including "Biotech in a Box" loaner equipment) designed and provided by the scientists or classroom visits by the researchers. The committee could not assess these programs because long-term tracking of their impact is not provided. NPGI researchers frequently tap into existing education and training programs on their campuses (NSTC 2001). Although better public outreach is needed, many PIs are not trained in K-12 education, and they cannot devote much time to it because of the demanding schedule of the research profession. To resolve this issue, some NPGI-funded programs hired a full-time coordinator who provides cohesive leadership for all their outreach activities (NSTC 2004). The committee enthusiastically endorses this concept and concluded that NPGI has set an example for other federal programs by appointing a professorial or an affiliate faculty-level education coordinator for each of its Coordinated Agricultural Projects (Interagency Working Group on Plant Genomes, personal communication, September 18, 2007).

Some of the NPGI-associated outreach initiatives have been remarkably large scale. The Plant Genomics Research Experience for Teachers at the University of Missouri has trained 70 teachers over the last four years (NSTC 2007). By creating educational software and online pedagogical materials, holding workshops, and providing equipment loans and ongoing support for teachers serving low-income, rural, and underrepresented minority students, the Partnership for Plant Genomics Education at the University of California, Davis, trained 52 teachers in FY 2005. They were expected to share their information with 772 other teachers and use activities and laboratories from the course with 8,600 students (NSTC 2006). Other K-12 outreach activities that could have broad impacts are listed in Appendix H.

Supplemental funding for the MaizeGDB enabled the creation of a central online repository that compiles links to outreach resources in one location, the Plant Genome Research Outreach Portal, or PGROP (PGROP 2006). The portal, which has pull-down menus, allows users to conduct searches by user type (high school teachers, undergraduate students, growers, public at large), plant species, topic (for example, proteomics), or resource type (for example, Web-accessible teaching materials and fellowships). One of the strengths of the gateway's interface is its capacity for directors of individual outreach or educational programs to upload information about their own programs (Baran et al. 2004). Although navigation of the Web interface is straightforward, searches routinely yield an unwieldy number of marginally relevant "hits." A search for "Resources for High School Teachers," for example, returns links to 130 resources including many Web pages on single genera of plants that distract from the relatively few resources (such as animated tutorials for use in the classroom) that are truly targeted specifically to teachers. On the other hand, searches for graduate programs or summer internships in plant genomics yield incomplete lists and nonfunctional links. PGROP is a comprehensive resource with high potential impact that would benefit from more inclusive cataloging and more robust, discriminating search functions.

International Interactions

Another successful aspect of the NPGI-funded efforts is their collaboration with international partners. The coordination among researchers from six groups across three continents in the *Arabidopsis* sequencing project (The Arabidopsis Genome Initiative 2000) paved the way for subsequent multinational endeavors (Table 2-7). Participating non-U.S. scientists in each of these projects are supported by their respective national research funding programs. The projects are overseen and coordinated by an international committee of scientists, typically elected by the research community. Such projects leverage the resources, expertise, and facilities of many countries to achieve a much richer and more comprehensive set of genome datasets than could be obtained by any single national effort.

The free exchange of information engendered by such collaboration maximizes efficiency and minimizes the duplication of efforts among teams of researchers. U.S.-funded projects, from *Arabidopsis* Genome and *Arabidopsis* 2010, through the entire spectrum of NPGI projects, have led the way in truly open access data deposition. Policy recommendations for U.S. funding must be fully self-contained, both intellectually and technically. While international collaboration is important, the success of NPGI and other U.S. science cannot be reliant on access to data and resources that, to date, are often only available with intellectual property strings attached.

A prominent example of a successful NPGI-supported international collaborative effort is the International Rice Genome Sequencing Project (IRGSP), a consortium of publicly funded laboratories from the United States, Japan, China, Taiwan, India, the Republic of Korea, Brazil, Thailand, and the United Kingdom. Two companies, Monsanto and Syngenta, invested in rice genome sequencing independently and their willingness to release data publicly facilitated the completion of the draft sequence, which was announced in 2002 (IRGSP 2002). The sharing of data, materials, and technology between public and private sector players hastened the completion of the projected 10-year initiative by four years.

Building on the success of the rice genome sequencing project, an International Rice Functional Genomics Consortium was convened with leaders from 18 institutions representing 10 countries and two international agricultural research centers. The goals of the initiative are to work cooperatively to elucidate gene

TABLE 2-7 Examples of NPGI-funded Projects That Involve International Collaboration

Project Name	Website
International Barley Sequencing Consortium	*http://www.public.iastate.edu/~imagefpc/IBSC%20Webpage/IBSC%20Template-home.html*
International *Brachypodium* Initiative	*http://www.brachypodium.org/*
International Citrus Genome Consortium	*http://int-citrusgenomics.org/*
International Cotton Genome Initiative	*http://icgi.tamu.edu/*
International Grape Genome Program	*http://www.vitaceae.org/*
International Legume Database & Info System	*http://www.ildis.org/*
International *Populus* Genome Consortium	*http://www.ornl.gov/sci/ipgc/*
International Rice Functional Genomics Consortium	*http://www.iris.irri.org:8080/IRFGC/*
International Rice Genome Sequencing Project	*http://rgp.dna.affrc.go.jp/IRGSP/*
International Soybean Genome Consortium	*http://genome.purdue.edu/isgc/index.shtml*
International Tomato Sequencing Project	*http://www.sgn.cornell.edu/about/tomato_sequencing.pl*
International Wheat Genome Sequencing Consortium	*http://www.wheatgenome.org/*
The Multinational Coordinated *Arabidopsis thaliana* Functional Genomics Project	*http://www.arabidopsis.org/* *http://www.arabidopsis.org/portals/masc/index.jsp*
Multinational Brassica Genome Project	*http://www.brassica.info/*
SOL Genomics Network	*http://www.sgn.cornell.edu/*
SOL (EU-SOL)	*http://www.eu-sol.net/*
SOL (Lat-SOL)	*http://cnia.inta.gov.ar/lat-sol/*

SOURCE: Interagency Working Group on Plant Genomes.

function, integrate databases, establish bilateral or multilateral partnerships, and enhance rice production (IRFGC 2007). NPGI-funded PIs are prominent on the project's steering committee, and the USDA Cooperative State Research, Education, and Extension Service (USDA-CSREES) has facilitated participation by American students, postdoctoral fellows, and senior researchers.

Other functional genomics projects also capitalize on the resources and expertise of an international scientific community. For example, the goal of the International Solanaceae Genomics is to develop a comparative framework for studying plant diversification and adaptation across the Solanaceae family (including the important crop plants tomato, potato, eggplant, and pepper). The SOL Genomics Networks form partnerships with laboratories in Latin America and in Europe to improve the nutritional value, taste, flavor, fragrance, shelf-life, starch composition, yield, and other traits important to consumers, producers, and processors of these staple fruits and vegetables (European Commission 2006).

The Developing Country Collaborations in Plant Genome Research (DCC-PGR) program was started as an NPGI activity in 2004 to support collaborative research involving researchers in the United States and scientists in developing countries. The goal is to facilitate the application of new tools and resources to solve agricultural, environmental, and energy problems of significance to the foreign researcher's home country. Supplemental funding to an existing or a new NPGI award of up to $100,000 for two years enables joint research projects and long- or short-term reciprocal visits of students and senior investigators, which could lead to long-term partnerships (NSF 2007b). International collaborative NPGI projects that are targeted to directly benefit resource-poor farmers in developing countries include the following (NSTC 2004, 2005):

- Using the genome map of sorghum, an important staple cereal in Africa and India, to elucidate networks of genes that control drought tolerance.
- Developing cultivars of the African cow pea (a legume widely grown in Africa, Latin America, Southeast Asia, and the southern United States) that are resistant to the parasitic weed *Striga*.
- Establishing comparative markers to link the genetic maps of chick pea, cow pea, and pigeon pea to the *Medicago* genome sequence map, enabling breeders in India and Africa to identify disease resistance genes and develop improved cultivars of their local crops.
- Using proteomics technologies to develop improved oilseed cultivars in Nepal with enhanced processing and feed characteristics.
- Harnessing genetic variation in natural rice populations to introduce disease resistance and drought tolerance from natural populations into improved cultivars.
- Developing new Bolivian cultivars of potato that are resistant to bacterial wilt, which causes serious crop losses each year.
- Investigating the genes that allow plants to produce seed without fertilization (apomixis), which can be used to breed desirable traits into land races of corn that are adapted to the diverse growing conditions across Mexico.

Two other major NPGI projects that involve substantial international collaborations and represent the next wave of genomics initiatives with applications to the developing world are the sequencing of cassava and the Generation Challenge Program. U.S. researchers and their partners at the International Center for Tropical Agriculture (a center of the Consultative Group on International Agricultural Research, CGIAR) are working with JGI to perform sample sequencing of the cassava (*Manihot esculenta*) genome. The tuber grows in diverse climates and in nutrient-poor soil and is an important source of food and biofuel for 1 billion people globally. As a staple for subsistence farmers, a cash crop for local markets, and a reliable source of food and animal feed in famines, *M. esculenta* is well positioned for nutritional improvement, but genome sequencing will also provide insights into starch and protein biosynthesis and stress controls (JGI 2006). The CGIAR Generation Challenge Program is dedicated to alleviating constraints in agricultural productivity that contribute to global poverty and hunger, with an emphasis on harnessing genomic technologies to make rapid progress in the area of drought tolerance (CGIAR Genomics Task Force 2006).

NPGI AND INTERAGENCY COOPERATION

Earlier sections in this chapter assessed various specific aspects of NPGI, but NPGI is not merely a funding mechanism. It is an interagency collaboration that coordinates activities in plant genomics. The committee also assessed the role of IWG in facilitating research, training, and outreach. In addition, perhaps the most important metric for the success of NPGI is whether, and to what extent, U.S. research and development agencies have reprioritized their mission-oriented, agency-specific research portfolios on the basis of NPGI research and discoveries.

Coordination of Programs

Although each member agency of IWG has its own mission, some agencies also have overlapping interests and goals. IWG member agencies have increasingly issued joint calls for proposals or co-funded programs of mutual interest. (See Appendix I for examples). The joint programs reduce administrative burdens for principal investigators applying for funds and allow the agencies to jointly achieve common program goals.

Perhaps the most important metric for NPGI is whether the science funded to date has served as a springboard for agency-specific, mission-oriented programs that capitalize on either new funding from the public or on public-private partnerships. One of the greatest challenges that the nation faces in the 21st century is reducing dependence on foreign oil. Top quality basic genome-based research is necessary to achieve these goals, and that research will rely heavily on plant genomics and genetics for its progress. Biofuels and bio-based products are potentially sustainable solutions if conversion efficiency is improved dramatically (NRC 2005a).

New investments in bioenergy research will leverage basic plant genomics discoveries made though NPGI (Table 2-8). The largest to date is the $500 million investment by BP America, Inc., in conjunction with the University of California, Berkeley, and Lawrence Berkeley National Laboratory to create the Energy Bioscience Institute (EBI). In addition, DOE has made plant genomics a linchpin in its

Genomics: GTL portfolio for Bioenergy. Using the JGI sequencing platform as a departure point, DOE recently invested about $375 million into three bioenergy centers that are intended to accelerate basic research in the development of cellulosic ethanol and other biofuels (*http://genomicsgtl.energy.gov/centers/*). Further, a joint DOE and USDA-CSREES program recently announced $8.3 million in grants for improvement of feedstock (DOE 2007a).

Additional new programs that leverage the basic science of NPGI include a DOE Genomics: GTL program that is soliciting proposals for new analytical and imaging technologies for lignocellulosic material degradation and for multiplexed screening for mutant plant phenotypes. Also, the 2007 Farm Bill Title VII (H.R. 2419) passed by the House of Representatives has provisions of $50 million per year for an Agricultural Bioenergy and Biobased Products Research Initiative and $100 million per year for a Specialty Crop Research Initiative to develop and disseminate science-based tools, including plant breeding, genetics, and genomics, to address needs of specialty crops (Table 2-8). At this writing, the bill has been placed on the Senate calendar for consideration. If it is passed, the initiatives would provide vital new resources to expand IWG activities.

Other examples of refocusing and increased investment in agency mission-specific research include the conversion of the USDA-CSREES National Research Initiative (NRI) Plant Genome panel into a translational genomics program in recent years, with a different crop focus each year, and with a major emphasis on outreach and extension of research efforts. USDA-ARS has also refocused some of its internal programs to complement and support NPGI research. The National Program 301 on Plant Genetic Resources, Genomics, and Genetics Improvement redirected its statement of purpose to support the new discoveries made by NPGI-funded research ($140 million in FY 2007; Table 2-8). As NPGI research generates valuable data, the need for database stewardship and informatics tools to use the data effectively becomes apparent. Therefore, the National Program 301 includes a component on crop informatics, genomics, and genetic analyses that addresses genome database stewardship and informatics development, structural comparison and analysis of crop genomes, and genetic analyses and mapping of important traits.

Likewise, the National Program 302 on plant biological and molecular processes has redirected its focus to applications of genomics to crop plants because of NPGI discoveries. As a result of NPGI-funded research on model plants, National Program 302 has refocused its objective to take advantage of the new genomic information and to advance it from model plants to crop plants ($40 million in FY 2007; Table 2-8). The goal is to translate plant genomics into crop improvement.

Applied mission-oriented, agency-based forest tree genomics programs have also been derived from basic discoveries made through NPGI. For example, a

TABLE 2-8 Examples of Agency-specific And Mission-focused Programs That Have Spun Off of, or Benefit from, Results of NPGI Research

Programs	Program Budget (in millions)
Energy Biosciences Institute, UC Berkeley, BP, LNL	$500 over 10 years
Agricultural Bioenergy and Biobased Products Initiative[a]	$250 over 5 years
Barley Coordinated Agricultural Program	$5 over 4 years
Bioenergy Research Center[b]	$375 over 5 years
Conifer Coordinated Agricultural Program	$6 over 4 years
National Program 301: Plant Genetic Resources, Genomics, and Genetics Improvement	$140 in FY 2007
National Program 302: Plant Biological and Molecular Processes	$40 in FY 2007
Plant Feedstocks Genomics for Bioenergy	$8 over 3 years
Rice Coordinated Agricultural Program	$5 over 4 years
Specialty Crop Research Initiative[a]	$500 over 5 years
Wheat Coordinated Agricultural Program	$5 over 4 years

[a]The Agricultural Bioenergy and Biobased Products Initiative and the Specialty Crop Research Initiative were proposed in the 2007 Farm Bill, which has not been passed by Congress at the time this report was written.

[b]The program budget presented for the Bioenergy Research Center includes funding for plant and microbial research and technology development.

multimillion-dollar Coordinated Agricultural Project from USDA-CSREES and USDA Forest Service (USFS) on conifer genomics began in 2007 and will allow association genetic studies of trees in the major breeding programs throughout the United States. **Each of the above examples is testament to the power of federal investment in competitive, peer-reviewed, curiosity-driven basic plant genomics research, and illustrates the return reaped in translation to agency-specific, mission-oriented applied plant genomics.**

In-kind Support and Distribution of Resources

Although some IWG member agencies fund plant genomics research, all of them contribute to the goals of NPGI by providing in-kind support, distributing resources, and keeping each other abreast of latest genomic technologies.

In-kind Support (provided largely by IWG member Agencies)

USDA-ARS

USDA-ARS funding of $3.6 million in FY 2002 and $8.5 million in FY 2006 for plant bioinformatics includes support for the following projects:

• The maize genetics and genomics database. This project aims to synthesize, display, and provide access to maize genomics and genetics data for the research and user communities.

• Identification of functional sequence in plant genomes through bioinformatic, genomic, and genetic approaches. This project aims to provide resources to characterize, track, and identify sequence associated with agronomically important traits.

• An integrated database and bioinformatics resource for small grains. This project aims to integrate small grains genetic and genomic data within the GrainsGenes database and link to relevant external databases. It also aims to develop software and interfaces to enhance utility for researchers.

• Curation and development of the Soybean Breeder's Toolbox and its integration with other plant genome databases. This project aims to implement web-accessible computation and visualization tools to enable comparison and transfer of agronomically important genetic information among soybean and other related species. The project also involves the curation and enhancement of the SoyBase and the Soybean Breeder's Toolbox and the coordination of the assembly and annotation of soybean whole-genome sequence.

DOE

In addition to funding individual research projects, DOE's contributions to NPGI include sequencing of plant species through its Community Sequencing Program (CSP) or Laboratory Science Programs (LSP). Examples of plant genome sequencing by DOE's Joint Genome Institute through CSP include *Physcomitrella* in 2005; *Selaginella*, sorghum, *Arabidopsis lyrata*, *Capsella*, *Mimulus*, and the chloroplast of *Campanulales* in 2006; and *Brachypodium*, *Aquilegia*, *Gossypium*, cassava, maize, soybean, and *Eucalyptus* in 2007. JGI was the lead organization in the sequencing of poplar (Table 2-1).

JGI is committed to plant EST sampling as well. For example, JGI agreed to produce ESTs for switchgrass and peach in 2007, as well as for eucalyptus, foxtail millet, and conifers—loblolly pine and 22 other species selected for their commercial and ecological importance or their ability to provide phylogenetic insight into conifer genome evolution—in 2008 (DOE 2007a) (see Table 2-1).

The committee notes that JGI's contribution to plant genomics is unique and fundamental, and spans both explicitly energy-oriented projects and projects that broadly inform all of plant biology from evolution through comparative genomics. There is no other high-throughput sequencing facility interested in serving plant genomics that can match JGI's power and consequent economy of

scale. These points inform one of the committee's most important recommendations (see Chapter 3).

USDA Forest Service

The USFS has 10 full-time-equivalent scientists who conduct genomics research. USFS has also provided technical support in tree genomics or molecular genetics in the form of competitive awards or cooperative agreements (see Appendix K). Compared to funding from USDA-CSREES NRI and NSF grant programs, USFS has to date made modest investments in plant genomics. Other in-kind products of USFS include:

- Maps of amplified fragment length polymorphism and single sequence repeats for the American beech.
- Markers for the selection of butternut that is resistant to butternut canker.
- Markers for the improvement of black walnut.
- Multiplex sequencing capability on high-capacity sequencing platform.
- Specific markers for identifying rust-resistant loblolly pine.
- Neutral markers for QTL analyses of loblolly pine.

NHGRI

Although NHGRI's primary focus is human genome sequencing, it plays a role in advancing NPGI's objectives through its support for genome sequencing and its built genomics infrastructure. NHGRI has provided financial support for a number of large-scale sequencing centers over the years. Although NHGRI does not fund plant genome sequencing directly, parts of some plant genome sequencing projects have been done at one of the NHGRI-supported sequencing centers, and many of the fungal pathogen genome sequences noted above were done as part of the Broad Institute's Fungal Genomics Program. NHGRI also supports the advancement of sequencing technology, development of bioinformatics tools, and identification of all functional elements in the human genome. As a member of NPGI, NHGRI can pass on the technologies and tools developed and lessons learned to the plant community swiftly.

NHGRI continues to promote free and open data release and keeps NPGI updated on NHGRI's policies. In fact, NPGI has adopted the Bermuda accord that requires rapid release of publicly-funded sequence assemblies of 2kb or larger and the Fort Lauderdale accord that defines a community-resource project. NHGRI considers whether the data release policies are appropriate periodically and keep NPGI informed on those discussions.

Distribution of Resources

The National Plant Germplasm System (NPGS), managed and funded by USDA-ARS in partnership with agricultural experiment stations and land-grant universities, aids plant scientists by conserving the plants and seeds of nearly 10,000 species. To ensure that genes are available to NPGI fundees, NPGS continues to acquire, preserve, evaluate, document, and distribute crop germplasms, many of which originate outside the United States (ARS 2005). NPGS distributed over 150,000 accessions in 2006, including 9,131 *Triticum*, 5,597 *Oryza*, 11,951 *Zea mays*, 19,349 *Glycine*, 9,729 *Lycopersicon*, and 5,073 *Vitus*. Among those distributions, some were mutants or cytogenetic stocks (based on information submitted to the committee by USDA-ARS on May 17, 2007). Because of the increasing demand as a result of NPGI-funded research, stock centers were built or expanded. For example, the Maize Genetics Corporation was expanded to provide long-term curation of maize mutant genetic stocks developed by NPGI awardees. The Genetic Stocks—Oryza Collection was established as a result of NPGI when the rice genome was sequenced and the need for a collection of rice seed mutant genetic stocks was recognized. Mutant seed genetic stocks of other plants developed by NPGI awardees are added to the working collections of other NPGS repositories. Other than germplasm collections, many Websites and databases were developed or expanded as a result of NPGI (see Appendix F).

3

Recommendations and Goals: New Horizons in Plant Genomics

"Our greatest responsibility is to be good ancestors." Dr. Jonas Salk

THE FUTURE OF PLANT GENOME RESEARCH

Plant science today lies at the nexus of potential solutions for global problems that are challenging a human population of more than 6 billion people today and that is projected to reach 9 billion by 2054 (United Nations 1999). Plants are extremely important sources of food, fiber, energy, and animal feed, yet plant biologists are only beginning to understand the fundamental principles of how plants grow and develop; how they cope with daily, seasonal, biotic and abiotic changes in their environment; how they participate in complex communities in diverse ecosystems; and how they evolved. Provision of adequate food and nutrition, expanded alternative energy sources, and sustainable environmental stewardship will require the development of new technologies for agricultural solutions that rest on detailed scientific knowledge. An understanding of the principles underlying plant growth, development, and reproduction will enable scientists to play a role in securing global health, the global economy, and the global environment by providing new options for improving productivity and reducing the environmental footprint of agriculture. The key to understanding those principles is basic research done in the context of the revolution of genome-based science.

The committee strongly recommends that the next wave of National Plant Genome Initiative (NPGI) research should have as its top priority innovative, competitive peer-reviewed basic science aimed at detailed and system-wide understanding of the functions of individual genes, how those functions are

connected in networks, and how they control plant growth, form, function, performance, and evolution. The last 10 years have witnessed an explosion of knowledge regarding the various individual pathways that control plant growth and development. Biologists now better understand the principles underlying how plants perceive changes in their ambient environment; how they respond to pathogens; how they build flowers, leaves, and roots; and how various classes of hormone receptors direct plant growth. Several plant genomes have been sequenced, a few of which were sequenced to high quality. These discoveries, coupled with continued genome sequencing and resequencing, are the springboard for the next 10 to 20 years, a time during which fundamental research would have the definition of a plant that is more than "the sum of the parts" as its goal.

Because of the federal research and development investments made over the last 20 years, plant biology is at the doorstep of an era of unprecedented large dataset collection, systems-wide analyses of those data, model building, and ever more precise hypothesis testing. The fruits of this research will be deeper understanding of how plant genomes condition important traits. However, the current knowledge is simply too underdeveloped, and translation of that knowledge is too costly or too imprecise, for the majority of desired applications. **Thus, NPGI should aim to produce knowledge and tools for efficient trait modification and technology leaps so that genomic information can be translated effectively into environmentally sustainable products of benefit to humankind.**

The committee recommends the following guiding principles to achieve those goals.

- **The committee strongly endorses the conclusions of the 2002 NRC report, *The National Plant Genome Initiative: Objectives for 2003–2008*, that studies aimed at defining core concepts of molecular and developmental plant biology are best undertaken rapidly and efficiently in model plant systems. Basic discovery that can be most rapidly and efficiently done in these systems should receive high priority.** The committee advocates deep investment in the broadest possible set of genomics tools for these carefully selected systems. These systems would be chosen on the basis that they can provide vital paradigms that inform many other aspects of NPGI and can maximally leverage continued, independent investments in *Arabidopsis* genome science.
- **Because the diversity of plant form and function utilized by humans is very broad, the committee strongly endorses the approach that parts of the overall genomics toolkit be deployed to investigate specific aspects of plant tissue and organ development, environmental adaptations, or biochemical processes that are not well represented in core model species.** This will include a great deal of genome sequencing along the entire plant phylogeny to inform comparative func-

tional studies. However, descriptive functional studies aimed at gathering parallel datasets merely because they are derived from crop species would not be a good use of resources and are best avoided.

- **The committee recognizes the critical need for the development and deployment of field-robust, high-resolution genotyping and phenotyping methods for use in molecular-assisted plant breeding across a broad swath of crops.** These methods will require DNA sequencing (though certainly not always full genome sequencing) and substantial population sampling to define informative markers. They will require technological breakthroughs at genotype and phenotype levels to produce simple, robust methods available to plant breeders in the United States and around the world. These activities are crucial if the ultimate benefit of the NPGI discovery engine is to be realized. Hence, a scaffold of genomic tools is needed in each of the major crops in order to translate model organism concepts to them.
- **The committee suggests that the priorities for NPGI and associated plant sciences be framed towards addressing the large challenges facing humanity, including bioenergy, climate change, sustainability, and human nutrition.** The committee envisions the growing enablement of genomic tools, systems biology, and trait modification capabilities in a wider range of species than those currently emphasized. However, investments in those tools are only justified when there is a clear social goal and when the technologies for data collection, hypothesis testing, and trait modification become reasonably efficient and robust.
- **The committee's nine recommendations for NPGI priorities in the future are listed in Box 3-1.**

Each recommendation has a set of goals on three different time horizons: The 5-year goals represent immediate, pragmatic "next steps" in plant genome science, 10-year goals require significant development of new tools and resources to enable transformative solutions to real world problems, and 20-year "achievements" reflect the committee's desire to define some admittedly long-range, high-risk, high-reward areas that would significantly alter society's ability to understand how plants work.

TOOLS FOR PLANT GENOME RESEARCH IN THE 21ST CENTURY

One of the most remarkable impacts of genomics projects is the development and application of facile technologies that allow the global analysis of cellular components, including genes, proteins, and metabolites. After their invention, high-impact technologies are disseminated for use by individual laboratories and by "data production centers" that generate large amounts of data to benefit the entire scientific community. The number of hypothesis-driven, single investigator

BOX 3-1
Overall Recommendations

RECOMMENDATION 1: Expand plant genome sequencing, plant-associated microbial sequencing, and plant-associated metagenome sequencing, and associated high quality annotation by (a) using the Department of Energy's Joint Genome Institute's sequencing capacity to generally serve plant sciences and (b) empowering individual principal investigators or collaborative groups to access and utilize next generation sequencing technologies for a broad spectrum of genomics and metagenomics discovery.

RECOMMENDATION 2: Develop "omics" resources and toolkits at high resolution in a few, carefully chosen plant species, including expansion and deeper investment in currently leading model species.

RECOMMENDATION 3: Develop "omics" resources at a broader, shallower level across a number of additional species to (a) expand the phylogenetic scope of functional inference, particularly when this is justified to test clearly specified hypotheses, (b) understand physiological and developmental processes to a depth that is not feasible in the model systems, and (c) provide the foundation to improve U.S. competitiveness of important crop and tree species.

RECOMMENDATION 4: Use systems-level approaches to understand plant growth and development in controlled and relevant environments, with the goal to create the iPlant, a large family of mathematical models that generate computable plants genuinely predictive of plant system behavior under a range of environmental conditions.

RECOMMENDATION 5: Increase the understanding of plant evolution, domestication, and performance in various ecological settings via investment in comparative genomics, and in the metagenomics of living communities of interacting organisms.

RECOMMENDATION 6: Enable translation of basic plant genomics towards sustainable deliverables in the field, and continue to use NPGI as a foundation for new, agency-specific, mission-oriented plant improvement programs.

RECOMMENDATION 7: Develop and deploy sustainable, adaptable, interoperable, accessible, and evolvable computational tools to support and enhance Recommendations 1–6.

RECOMMENDATION 8: Improve the recruitment of the best, broadly trained scientists into plant sciences.

RECOMMENDATION 9: Promote outreach on plant genomics and related issues that are critical to educating the American public on the value of genomics-based innovations.

research projects will have to grow so that creative scientists can avail themselves of these technologies and capture the resultant benefits for society. It is equally imperative that groups of investigators, whether within or across institutions, be supported for collaborative projects when they are scientifically warranted. Collaborative group formation, however, should not be a requirement for funding because these can be "forced marriages of convenience" that are often not synergistic in their output. Finally, the plant genomics community has benefited from the establishment of high-throughput production centers and will continue to do so. These are particularly well suited to generation of data and resources for use by the broader community.

In principle, production centers that produce physical or information resources have the advantages of higher efficiency and uniform quality control standards to ensure that useful reagents and information are produced. The guiding principle would be that the quality of information produced by a resource center be equal to or greater than that typically produced by an individual research laboratory.

Examples of genomics technologies that have had significant impact on the plant biology community include T-DNA and transposon tagging strategies, DNA microarrays, and mass spectrometry. These technologies are now sufficiently widespread that they are accessible to most researchers for individual experiments. In plant sciences, the accessibility of these technologies can be largely attributed to NPGI and the *Arabidopsis* 2010 Project of the National Science Foundation (NSF). At the same time, these technologies also are used in production projects. For example, the ends of T-DNA and transposon insertions are sequenced to locate the position of each in the genome. TILLING collections now exist for various species, and they allow investigators to screen for point mutations using polymerase chain reaction. DNA microarrays are used for large-scale analysis of gene expression and mapping transcription factor binding sites. Mass spectrometry is readily used for large-scale mapping of protein-protein interactions.

DNA Sequencing: The Basis of Genomics

Recognizing and taking advantage of opportunities to "upgrade" large-scale datasets as new, quantitative, rapid, and cost-effective technologies are released is critical to NPGI. It is also important that NPGI lead the development of such technologies, which would then drive their deployment via the mission-based member agencies like the U.S. Department of Agriculture (USDA) and the U.S. Department of Energy (DOE). An example of opportunities for "data upgrades" is the new, high-throughput next-generation DNA sequencing technologies that have emerged in the last year—for example, pyrophosphate sequencing (454 Life Sciences™/Roche) and localized cluster sequencing (Illumina, Inc.). Others will no doubt emerge very soon.

The new technologies grew out of a specific funding mechanism from the National Human Genome Research Institute (NHGRI) to support the goal of sequencing a human genome for $1,000. The technologies have had enormous impact on the sequencing of new genomes and have profoundly altered the ability to resequence, at a huge savings, natural variants of species where a reference genome sequence already exists. The new, next-generation DNA sequencing technologies will revolutionize the ability to map transcribed regions and transcription factor binding sites across a genome and to address how these phenotypes change over developmental time and in response to various stresses.

Resequencing technologies can open new vistas in creative analysis of natural variation and evolution, and in understanding the complexity of organisms present in environmental samples of plants and their associated microorganisms. In turn, next- generation DNA sequencing technologies have created demand for new informatics tools that can deal with the collection and assembly of small DNA fragments. This interplay results in a familiar and compelling cycle—important new technologies drive the creation of new ancillary technologies and create horizons for new biological experimentation that were previously unreachable, leading to new levels of detailed experimental understanding.

Thus, the committee can now credibly propose to use genomics to understand the principles underlying plant genome structure and evolution. Understanding how plant genomes expand and contract through polyploidization, segmental duplication, and subsequent loss or silencing of genetic information, for example, is now within reach. Furthermore, what were once puzzling and unappreciated features of plant genomes, such as the very high proportion occupied by transposons in some lineages, can be understood within a solid theoretical framework with genome sequencing on the scale recommended in this report. The committee does not, however, anticipate that physical chromosome maps and complete draft sequences will be required for all projects. Judicious choices for genome sequencing, in addition to those species listed below in Table 3.1, should consider how polyploidization has led to variable plant gene function and the evolution of novel traits of interest.

RECOMMENDATION 1: Expand plant genome sequencing, plant-associated microbial sequencing, and plant-associated metagenome sequencing, and associated high quality annotation by (a) using the Department of Energy's Joint Genome Institute's sequencing capacity to generally serve plant sciences and (b) empowering individual principal investigators or collaborative groups to access and utilize next generation sequencing technologies for a broad spectrum of genomics and metagenomics discovery.

As noted in Chapter 1, genome sequence is the raw material for functional, evolutionary and translational tool development at the center of plant genome sciences. Plant sciences will benefit from the generation of the first "reference genome" sequences for a growing number of species that define key points in plant evolution. The next-generation sequencing will enable both "reference genome" sequencing and resequencing for purposes of population and evolutionary genomics (see also below). As an example, it is likely that sequences from closely related species will constitute a powerful way to inform the functional biology of target genomes (from patterns of evolutionary conservation of sequence motifs, functional domains, and so on). Hence, for every species whose genome is chosen for a reference sequencing project, parallel sequence analysis of a related taxon of appropriate evolutionary distance (something on the order of 30 to 50 percent divergence at silent sites being optimally informative) would be appropriate.

Furthermore, the metagenomes of cultivated plants and plants in natural ecosystem communities will provide rich arenas for future discovery of important interorganismal associations that have positive or negative impact on plant performance (NRC 2007a). Metagenomics has been embraced by NIH, and has led to a major program on the human metabiome. A similar large-scale investment in plant-associated metagenomics is justified because of the diversity of plant-associated microbial communities and their impact on plant productivity. For example, the communities of microorganisms associated with candidate perennial biofuels crops, in monoculture or in more natural assemblages, are not well understood. As another example, the rhizosphere community, both microbial and animal, can influence root growth and development. Certain microorganisms can protect plants from other pathogenic microorganisms. Hence, a merging of metagenomics with root genomics would be rewarding. Thus, the committee strongly endorses the recommendation that NPGI make major investments in both plant genome and large-scale metagenomics sequencing efforts.

The unique role played by the Department of Energy's Joint Genome Institute (JGI) in the service of NPGI is critical. Although there are several high-throughput genome centers devoted to the missions of NHGRI, only JGI has plant biology as a central component of its mission. JGI has established a peer-reviewed policy for high-impact reference plant genome sequencing, which it has implemented successfully (see Chapter 2). The economies of scale gained from JGI's expertise and throughput, especially with their addition of next-generation sequencing capabilities, is unlikely to be matched by another sequencing center that has *a deep interest in plant genomics.* JGI is thus uniquely placed for the development of projects that combine traditional Sanger sequencing with the next-generation sequencing technologies that will lower the costs of reference sequencing considerably and allow economies of scale for resequencing projects. Table 3-1 provides a list of species for which one could argue a strong case for inclusion for NPGI genome sequencing plans in the next 10 years of NPGI.

Even at the economies of scale provided by combining JGI's throughput and next-generation sequencing, the sequencing of one reference genome from each of the species listed above will be costly. Therefore, use of other criteria to prioritize the list is necessary. The committee's recommendation for criteria to prioritize organisms for sequencing, as applied to different sets of biological and technologic issues, can be found in Recommendations 2–4 below and in the 2002 NRC report.

JGI should also seek to upgrade its basic and limited annotations, preferably via collaboration with groups containing the relevant expertise or by expanding its own activities in this area. Interaction between JGI and the NSF's Plant Cyberinfrastructure awardees could be synergistic in this regard. The committee therefore considers it highly desirable that DOE continue to take a broad view of JGI's unique position in the plant science community. It is critical to the success of NPGI that JGI continue to serve a *broad* remit for sequencing and resequencing of plant genomes, a remit not limited to only the sequencing of plants that are directly important to bioenergy production. To narrow JGI's mission would imperil a successful pillar of the NPGI infrastructure.

The next-generation sequencing technologies and supporting bioinformatics will also make resequencing of many different genotypes of small genome species a reasonable goal for individual principal investigators (PIs) or for groups of PIs. Resequencing is a critical new tool in the genomics toolkit because it allows scientists to understand how individuals vary at the DNA level, and how that variation shapes differences between individuals of the same species, and across short evolutionary distances by sequencing individuals of closely related species. Resequencing is especially important in the context of understanding evolutionary mechanisms and the natural diversity of plant form and function. Resequencing is already having a powerful impact on *Arabidopsis* genomics (Clark et al. 2007; Kim et al. 2007), and a project underway to resequence many rice relatives will certainly have similar impact in the understanding of rice evolution and domestication.

It seems reasonable that the JGI would take the lead on generating a broad swath of new plant genome sequences, because plant science still requires many high-quality draft sequences to serve as reference sequences for those species and branches of the evolutionary tree. In addition, other existing large-scale sequencing centers could be recruited to participate in NPGI activities. The costs of sequencing will likely drop, and many of the major crop species could be sequenced. Furthermore, multiple reference sequences might be necessary to cover the major halpotypes of a given species, if the haplotypes are divergent enough from one another. By contrast, resequencing efforts could be done by individual laboratories with access to the new sequencing technologies, or consortia of investigators interested in specific questions in population, evolution, and ecology that require a large resequencing component. Indeed, the *Arabidopsis* Landsberg-*er* and Cvi-0 accessions have been resequenced in two weeks each at a fraction of the cost of the original reference Col-0 sequence (J. Ecker, Salk Institute, personal communication, October 20, 2007).

Goals for Sequencing (Recommendation 1)

5-year goals

- Sequence the genomes of 25–50 strategically chosen plants and resequence the genomes of hundreds, if not thousands, of wild accessions of the plants chosen for the full "omics" effort. These sequencing programs would be accompanied by standards-based annotation.

20-year achievements

- Hundreds of reference plant genomes will be draft sequenced to high coverage and annotated for comparative purposes and development of mapping tools. These will blanket the plant evolutionary scale.
- Tens of thousands of plant genomes, or more, will exist as annotated resequences.

"Omics" Resources and Toolkits

RECOMMENDATION 2: Develop "omics" resources and toolkits at high resolution in a few, carefully chosen plant species, including expansion and deeper investment in currently leading model species.

All well-planned genome initiatives involve systematic development of resources that enable next-generation experimentation. NPGI is no exception. These resources include tools for genomics, epigenomics, transcriptomics, proteomics, metabolomics—often referred to collectively as "omics" tools. The tools result from large datasets that, for example, catalog mRNAs or small RNAs, proteins, or metabolites. But they also result in experimental materials, such as mutant plants, cDNA clones, and recombinant proteins. Development of omics tools most commonly requires high-throughput, computationally intense methods, and it is technology-driven. As computation and technology advances, so do the quality and quantity of omics data and resources. The utility of omics tools depends on accessibility and applicability to a broad community of researchers.

TABLE 3-1 Desirable Reference Genome Sequences Not Currently Funded. This Table Lists Other Future Projects of Direct Relevance to Food, Feed, and Fuel Needs of the United States

Species	Common Name	Genome Size (Gb)	Notes
Phaseolus vulgaris	Common bean	0.5	An important crop in its own right, *Phaseolus* is also an unduplicated outgroup for the recent soybean tetraploidy.
Pinus taeda	Loblolly pine	20	Wood crop, forest resources. Other gymnosperms (for example, spruce) also desirable, but note extremely large genome size.
Pennisetum glaucum	Pearl millet	2.7	Drought tolerant grass, cereal of "last resort."
Panicum capillare	Diploid switch grass	0.5	A genetically tractable diploid relative of the tetraploid Panicum virgatum (switchgrass), a leading biofuel crop.
Triticum aestivum	Hexaploid bread wheat	17	Hexaploid wheat and its diploid relatives, which are major sources of nutrition around the world and a system for understanding genetic effects of domestication and polyploidy.
Aegilops speltoides, Triticum monococcum, Aegilops tauschii	Diploid wheats related to progenitors of bread wheat	2-4	
Malus x domestica	Apple	0.7	Along with peach, these two rosaceous crops are at strategic phylogenetic distances for intrafamily sequence comparisons
Fragaria vesca	Strawberry	0.2	
Musa acuminata	Banana, plantain	0.6	Outgroup for grasses and the grass-specific paleotetraploidy, and therefore key to understanding important crops, especially in developing world. Vulnerable through limited genetic diversity.
Citrus sinensis	Sweet orange	0.4	Major U.S. crop that is highly sensitive to frost. Genome sequencing could aid genetic improvement for cold resistance.
Marchantia polymorpha	Liverwort	0.4	Primitive land plant that will assist in understanding the polarization of changes along the stem leading to angiosperms and gymnosperms.
Manihot esculenta	Cassava	0.8	Source of carbohydrates in developing world. Sample sequencing project is underway, with no full genome commitment

TABLE 3-1 Continued

Species	Common Name	Genome Size (Gb)	Notes
Gossypium sp	Diploid cotton; polyploidy cottons	>1.0	Valuable fiber crop is polyploid, with diploid relatives. Sample sequencing project is underway, but no full genome commitment.
Saccharum officinarum, Miscanthus sinensis	Sugarcane, Chinese silver grass	2-3	Rapidly growing C4 grasses with potential for biofuel feedstocks. Sugarcane is octoploid, *Miscanthus* is diploid, also providing a rich system for studying polyploidy.
Citrullus vulgarus	Watermelon	0.5	Would provide a cost-effective reference genome for cucurbits
Lactuca sp.	Lettuce	2.3	Diverse complex of species provides rich gene pool for breeding hardier varieties.
Solanum tuberosum	Potato	0.9	Although related to tomato, potatoes were independently domesticated. Important comparators within Solanaceae.
Solanum chacoensis	Wild potato	0.6	
Ipomoea sp.	Morning glory	0.7-1	Morning glory, diploid closely related to sweet potatos, which are typically polyploid. Potential genetic model system for tuber formation.
Helianthus anuus	Sunflower	2.4	Important source of edible oil worldwide
Antirrhinum majus	Snapdragon	1.6	Genetic model system that would be invigorated by genomic resources
Medicago sativa	Alfalfa	0.9	Forage crop, tetraploid relative to *M. truncatula* model system
Boechera holboellii	Rockcress	0.2	Model system for asexual (apomictic) reproduction in plants, with an international user community. Closely related to *Arabidopsis*.

Useful, integrative, Web-based computational resources that allow the broader community of scientists to derive high value and to form testable hypotheses are a critical component of a full omics effort. For example, what good is an omics project to assemble a deep catalog of molecular and metabolic responses to drought stress if plant biologists working on important problems of drought stress cannot access, synthesize, understand, and analyze the data? Furthermore, these Web-based computational tools should integrate as many omics resources as possible. Integration of distinct datasets, originating from different laboratories and production centers, is essential to understand the whole plant system.

Although the technology to derive omics data can be applied to a broad range of plants, including crop species, the extent to which value can be extracted from the resulting data and resources depends on a number of factors. **The criteria that enable the most productive generation and utilization of omics resources have not changed significantly since their articulation in the 2002 NRC report. However, the decrease in costs has certainly opened new vistas for some resequencing projects where a reference genome sequence now exists.**

Criteria for Investing in a "Full Omics Effort"

Priority would be given to "full omics" efforts to a small, select set of plant species, in which the most advanced resources and tools are developed and applied, and the data disseminated in a comprehensive format. For some resources, the full omics effort might involve invention or development of new technology. The full omics effort also involves redevelopment or refinement of resources as technology advances. In all cases, however, the tools and resources would serve as models, references, and guideposts for the broader plant science community. Maintaining a set of intensely studied full omics effort plant systems, especially through cutting-edge development of technological resources, is viewed by the committee as critical for subsequent discovery and application by a broad plant science community that extends beyond those working in core model systems. Detailed, and still relevant, criteria for full omics support were previously outlined in detail (NRC 2002). They are summarized here:

- **Complete, well-annotated genome sequence**. The importance of a high-quality, well-annotated genome sequence for development of the full omics toolkit cannot be overstated. The genome sequence provides the backbone resource on which all other resources depend. One cannot adequately interpret, for example, gene expression patterns from a microarray dataset without understanding where all the introns and exons occur. Similarly, epigenomic resources, such as maps of DNA methylation, cannot be assembled and placed in context without a properly annotated, complete genome sequence that identifies both protein-coding and noncoding genes.
- **Extensive bioinformatics resources**. Well-developed, community standards-based bioinformatics resources for a plant species are critical to place omics data in context. Community-supported resources, such as the *Arabidopsis* Information Resource (TAIR) or Gramene, serve as both clearinghouses for omics data and

sources for information required to develop omics tools. As omics data in model systems grow, so do the needs for better, more integrative bioinformatics and computation. Of particular importance in the future are computational methods, tools, and databases that integrate disparate omics data from multiple technology platforms and laboratories.

- **Traits of a "classic" model system.** Omics data, like most other types of scientific data, have to stand up to experimental validation and serve hypothesis-driven experimental science. Omics data have little value if robust experimental tools that meet community standards are not readily available. Therefore, omics tools and resources are best developed and tested in plant species that have served as model systems for basic experimental science. Broadly applicable model systems have several common features, including well-established and easily managed genetic properties (for example, the ability to self- and cross-fertilize, and abundant genetic markers); reasonably fast generation cycle times; small size; simple genetic transformation methods; well-documented protocols for biochemical, cell biological, and physiological assays; and easily accessed public resources, genetic stocks, and other experimental tools.
- **Ability to make "knockouts" or "knockdown" mutants.** Omics data allow researchers to form hypotheses about gene or protein functionality. Testing these hypotheses frequently requires analysis of mutants with defects in genes of interest. Public availability of indexed mutant collections (based on T-DNA insertions or transposon generated mutations), and more recently gene silencing collections, for the *Arabidopsis* and maize research communities have had an enormous positive impact on both the rate of progress and quality of the resulting science. Indeed, assembling and indexing of these mutant collections are themselves omics activities (functional genomics).
- **Genetic resources that allow definitions of gene and genome function across the range of available natural variation.** The domestication of plants from wild progenitors, and the ability to tap the breadth of ancestral genotypes as they are reflected in modern germplasm is, combined with next-generation DNA sequencing technologies, a new and powerful criterion for selection of a species for a full omics build out. See the specific list of criteria for such selection in Recommendation 3 later in this chapter.
- **A critical mass community of scientists.** Formation of high-quality omics resources and subsequent creative utilization requires a vibrant, enthusiastic community of scientists working on a common plant taxon. Again, the aggregating effect of a well-established plant model system promotes development and utilization of tools. Among the *Arabidopsis* worldwide community, for example, over 1,000 laboratories are listed on TAIR. Similarly, the list of maize "cooperators" at the Maize GDB approaches 3,000 individuals and organizations worldwide.

In view of these criteria, and in light of limits to financial resources, only a handful of plant species will warrant the investment to generate and exploit the full omics toolkit. As appropriate species are selected, they should represent major evolutionary branches within the plant kingdom (as described in the 2002 NRC report) and should leverage the productive investments in crop species made to date by NPGI and by the *Arabidopsis* 2010 Project.

RECOMMENDATION 3: Develop "omics" resources at a broader, shallower level across a number of additional species to (a) expand the phylogenetic scope of functional inference, particularly when this is justified to test clearly specified hypotheses, (b) understand physiological and developmental processes to a depth that is not feasible in the model systems, and (c) provide the foundation to improve U.S. competitiveness of important crop and tree species.

Criteria for Investing in a "Partial Omics Effort"

Although only a few species will fulfill the criteria for full omics treatment, a broad range of plant species will warrant development of a partial omics toolkit. There is not a strict, sanctioned list that defines which omics comprise a "partial omics toolkit," as the omics most relevant to a particular species or crop plant will vary. However, omics that are anticipated to have particularly broad relevance to many plants include transcriptomics (mRNA), small RNAomics, proteomics, and possibly epigenomics (for example, genome-wide DNA methylation patterns). Generally, partial omics plant species will have important practical, economic, or scientific features that are biologically distinct from those of intensely studied model plant species. Partial omics efforts enable concepts to emerge, and to be significantly extended, from model species through comparative approaches. This concept was discussed in detail earlier (NRC 2002); that discussion remains germane and can now be expanded to include more species because costs for many of the omics tools have dropped and will certainly continue to do so over the coming decade. Criteria for consideration for partial omics tool development include the following:

- **Important clade-specific processes and attributes.** Many plant species possess unique but important and interesting biological features that can be addressed using less than full omics approaches. Development of diverse fruits, formation, and properties of fiber and wood, and adaptation to fill broad niches or to tolerate extreme environments, are but a few examples of processes or attributes that cannot be studied sufficiently using core model systems.

- **Excellent comparative properties.** A major outcome of plant-based genomics over the last decade has been a revision of concepts about genome evolution. Many more questions can now be addressed using omics technology applied to close relatives of model and reference plants and to key points and transitions in plant evolution. In particular, high-throughput sequencing technology and computation can now provide vast datasets to increase scientists' understanding of natural variation, speciation, clade-specific adaptations, polyploidy, and many other evolutionary events.
- **Leveraging, adding value to a significant community, and making best use of existing resources.** NPGI has invested heavily in crop plant genomics. In many instances, the investments have provided omics resources that can now be exploited through translational genomics. Although often overlooked, the public-sector plant breeding community capacity for field phenotyping in crops such as wheat and soybean is substantial. However, access to cutting-edge field phenotyping and genotyping tools, some of which will require new technologies to be specifically developed—for example sophisticated remote sensing over time. In addition, national needs (for example, biofuels and specialty crops) are dictating both private and public research investments beyond the scope of NPGI funding. Getting the most out of current and previous productive investments needs to be considered in future omics investments of the NPGI.

In addition to the development of new resources, both the full and partial omics efforts will require the development of physical and information resources. Physical resources include items such as clone and mutant collections. Information resources include DNA sequences, datasets that catalog information such as protein-protein interactions, gene expression, and transcription factor binding. Those resources are extremely valuable for both individual researchers and those attempting to understand whole systems. A particularly relevant example is provided by the seed collections that are the backbone of mutant identification in any species. The committee notes that creation of user-friendly data interfaces for these physical resources requires a different skill set than data analysis.

Information resources are also particularly powerful for elucidating the emergent properties of systems. Global analysis of regulatory circuits in yeast has revealed logic concerning network organization and common regulatory motifs. Information resources are critical to modeling biological properties of plants. Such information is a prerequisite for proper prediction of biological properties and outcomes of plant manipulation.

Goals for "Omics" (Recommendations 2 and 3)

5-year goals

- Develop two to three centers for technology development, application to experiments, data management and analysis, and user interfaces. These centers should also curate the data generated by individual laboratories. These centers will serve broad communities of researchers.
- Develop full omics kits for a small set of model plant species that, via comparative approaches, can benefit a broad range of plants. Initially, the full omics toolkits would be built for *Arabidopsis* and a few other plants. This could be expanded as the price to conduct functional genomics decreases, and as the community resources required for a full omics effort are established in other species.
- Use the full omics toolkit to accelerate definition of gene function across the full omics model species and their sequenced closest relatives.
- Begin development of, or expand existing, partial omics kits for plant species that have important economic, ecological, developmental, or biological features that cannot be fully studied in the core model species. Some examples to consider: a legume (for nodulation studies), tomato (for fruit ripening), a major woody tree such as poplar, and additional high-value agriculture crops.
- Invest in the application and development of new methods to analyze gene sequences and allelic, genomic, and population-level DNA variation, gene expression, protein expression and interactions, and metabolite expression. Although the committee does not suggest that fully independent technologies be developed for plant research, it recognizes that some technology development might be necessary to adapt existing technologies currently used for animals and microorganisms to plant systems. This could begin with the full omics species and expand outward as costs drop and overall funding increases. The data obtained constitute the core information needed for developing an integrated systems approach, as outlined in the next section. Although the human genome project will develop many of these protocols, NPGI still needs to play an active role in technology development. The technologies include, but are not limited, to:
    - Methods to quantitatively monitor and analyze gene expression, including in specific cell types, at a greatly reduced cost.
    - Methods to critically characterize all protein-DNA interactions to identify regulatory sites on the genome of *Arabidopsis* and perhaps one additional model organism ENCODE (MOD-ENCODE) project. This should be enlarged as costs allow.
    - High-throughput production of affinity reagents to follow protein abundance, localization, and other applications.
    - Methods to monitor the expression of all proteins and other molecules (including low abundance) in a small cellular sample.

  - Methods to monitor all protein modifications in a small cellular sample.
  - Analyzing thousands of protein-protein interactions in parallel.
  - Analyzing large number of enzyme activities from diverse plant species, and at low cost.

10-year goals

- Implement high-throughput phenotyping and omics data collection under both controlled and field conditions.
- Understand the functional consequences of epigenomics.
  - Describe the full suite of changes to chromatin (histones, methylation) as a function of development and response to environment.
  - Define the genomic content of all small and ncRNAs.
  - Define functions for all miRNA and tasiRNAs (and all the others).
- Develop computational protocols for systematic integration of disparate omics datasets from different plant species.
- Create genome-wide knockouts and complete full-length cDNA collections; strive for accurate genome annotation, functional analysis and other omics tools for both core full omics models, and leading partial omics species.
- Implement translation of discoveries in model plants to most key crop plant species, including transformation methods for all major crop species.
- Use artificial chromosomes for rapid analysis and manipulation of complex traits.

20-year achievements

- In core model plants, omics data, high-resolution imaging, computation, and modeling will yield accurate, predictable mathematical models for major growth, signaling, and response pathways.
- Technology for de novo sequencing and resequencing, coupled with computing, will bring analysis of all plant species within the genomics umbrella.
- Technology will allow metaomics analysis of plants in communities, over single and multiple generations.

RECOMMENDATION 4: Use systems-level approaches to understand plant growth and development in controlled and relevant environments, with the goal to create the iPlant, a large family of mathematical models that generate computable plants genuinely predictive of plant system behavior under a range of environmental conditions.

NEW HORIZONS FOR PLANT GENOME RESEARCH

From Reductionism to Systems Biology to Modeling and Back

NPGI research to date has generated impressive advances with respect to generating DNA genome and mRNA sequences from basic model plants like *Arabidopsis* and key crop plant species like maize and rice, with more in process (Chapter 2). Further, the NSF *Arabidopsis* 2010 Project and the Interagency Working Group on Plant Genomes NPGI have defined an initial set of key pathway components required for the regulation and manipulation of plant growth and development and of plant responses to pathogens and environmental stress (see Chapter 2). As with any discipline, the reductionist experimental approach mostly creates sporadic information—a static biological "parts list." Critically, however, reductionist genetics-based data is explicitly linked to genetic causality, and thus typically results from this kind of research are associated with high confidence. The parts list is meant to reconstruct a dynamic view of the living plant that will then be used as the basis for crop improvement. Although the genome sequence of any given plant is a critical piece of information, it is not by itself a blueprint for understanding genome function.

The emergent properties of complex systems cannot be elucidated by examining the individual parts; rather, a systems-level approach is required to consider how many components act in concert. Gene-by-gene approaches, and even single network approaches, cannot lead to understanding of the diverse function(s) of every reference plant gene. To date about 20 percent of plant protein-encoding open reading frames—(ORF)eome genes—still have unknown functions. Over 40 percent of plant genes can be considered poorly characterized with respect to their function. These unknowns are the tip of the "anony-nome" iceberg, as the noncoding and small RNA genomic content are yet to be fully described, let alone functionally characterized. Considering that genes can have several functions when integrated over developmental time and in various environments, only a small fraction of plant genes can be said to be functionally characterized.

Representation of the regulatory plant needs to move beyond wall posters depicting two-dimensional metabolic pathways toward expansive, dynamic representations of plant processes that require multidimensional data visualization. Only with multidimensional visualization can the constantly switching circuitry used by plants while they are acting in their environments be observed with sufficient fidelity and resolution.

The systems biology approach recognizes that any given gene functions only in the context of particular *regulatory modules* made up of interacting *nodes* and *hubs* (other gene products and their interactions), and that each regulatory module is in constant flux with respect to others to form a *biological network*. The individual plant itself constitutes an organismal network of such component biological networks, and ultimately each organism participates in larger ecological networks made up of many types of organisms and their biotic and abiotic interactions. **The grand challenges toward achieving this level of characterization are**

- **A scalable view of global regulatory networks, from intracellular events to whole ecosystems, arrived at by large-scale collection of system-wide data sets from naturally variable genotypes (from genomes to phenomes) assessed across growth and stress conditions, and in association with other organisms like pathogens, mutualists, commensals, and symbionts.**
- **Families of models incorporating these datasets that both describe system behavior and predict outcomes of subsequent system perturbations.**
- **Validated computational representations of individual plant cells, tissues, and, eventually, whole plants interacting within their multiorganismal communities that enable accurate prediction of growth patterns and responses of plants under diverse natural environmental conditions.**

Characterizing regulatory networks in a systems manner that remains firmly mechanistically based is a major challenge for plant genomics over the next two decades. Networks define plant growth and development, and plant responses to environmental challenges. It will therefore be imperative to elucidate the networks underlying the traits of interest to scientists and breeders alike, in both controlled environments and under conditions relevant to their ecological and economic properties. These traits include growth and development through the life cycle, heterosis, abiotic stress tolerances, the control of flowering and fruit or seed production, the mechanisms of resistance to pathogens and herbivores, and symbioses with beneficial microbes, to name some of the most important.

Many of the basic intracellular algorithms can likely be determined from a few well-studied models; heterogeneous data collection and analysis will not be necessary for every plant species. As discussed in the section on omics toolkits, many of the most important basic lessons learned from the systems view of model species can be applied to other species. The plant research communities working on *Arabidopsis* and other well-developed species are beginning to accept and adopt the systems view and approach; incentives could be provided to the research communities to broaden these efforts.

However, as studies progress beyond basic plant biology and experimentally controlled environments and into specialized biological processes and uncontrolled environmental responses across the breadth of plant biology, the data from the models might fail to make useful predictions about higher level, emergent properties such as water and nutrient use efficiency, susceptibility to complex diseases, and yield. A major effort will, therefore, be needed to calibrate, and extend, beyond algorithms based on the elite model organisms. Selection of these species would be in alignment with the criteria for partial omics kits discussed above.

The long term output of systems-based research on regulatory networks should be a greatly enhanced ability to create custom plants highly tailored to meet a required need. Those needs could include predicting the capacity to produce adequate food and fuel under specific environmental challenges—for example, changes in pathogen loads or abiotic stresses associated with climate change.

Molecular Regulatory Networks

Plants might appear to the human eye to be "sessile," but they are just as dynamic in organizing cellular activities and sensing and responding to a constantly changing environment as animals. Indeed, plants' inability to move into and out of different environments has resulted in the evolution of remarkably powerful biosensing capabilities (for example, light quantity and quality, volatile molecules, presence of pathogens compared to symbionts and mutualists) that are necessary for their survival in a given niche. Although plant biologists have made important progress in defining components of the key regulatory modules and their hubs in plant species, a far more detailed and dynamic view is needed.

Biological networks have all evolved to be generally robust—that is, they are resistant to most stochastic changes they experience. However, the system robustness often comes at the price of fragility to unusual perturbations. The fragility can be exploited by the use of *perturbing factors* (for example, mutations, chemical treatments, stresses, and others) to probe how a network responds to the alteration of a given node. An understanding of the composition and regulation of all plant transcriptional networks is necessary to predict plant responses to perturbations. Plants have greatly expanded transcription factor families when compared to animal genomes, making straightforward loss-of-function genetic approaches less productive than systems approaches using omics tools. How the transcriptional networks are organized in a hierarchical fashion, the dynamics over multiple timescales (from seconds to seasons) of these networks, and how the transcriptional networks are layered with many other control modes (posttranslational regulation, protein dynamics and complex formation along with ion and metabolome fluxes and metabolic enzyme activity) need to be defined. Gene families encoding proteins responsible for regulated protein degradation are also expanded in plants, representing another layer of regulatory modules that needs to be fully elucidated. Given the recent development of mass spectrometry-based methods for profiling protease and kinase activities, these types of experiments are now within reach of the NPGI programs.

Preparing for the Application of Epigenomics to Crops

Plant biologists' knowledge of how epigenetic changes in the genome control critical plant processes is growing rapidly, but is still at a relatively early stage. For instance, from studying model systems, it is known that key developmental changes in plants can be controlled by small RNAs. The role that small RNAs and changes in the transcriptional status of chromatin play in gene expression is still largely a "black box" in crops. Given its relatively low cost, cataloging small RNA populations in several crop species could be an initial step towards understanding the mechanisms of epigenetic change across diverse crop species. It is also now feasible to use chromatin immunoprecipitation methods together with tiling arrays or deep sequencing to describe changes to histones and DNA methylation during development and response to environment in a variety of crop species.

Model Development, Iterative Experimental Refinement, and the iPlant

The large-scale data collections that are required to elucidate biological networks present a set of tough challenges to any research community. Although plant biologists can deposit omics data sets into well-organized databases, it is difficult to extract or predict emergent properties of a system from database formats. Such large data sets also overcome any intuitive ability to build component interactions into networks.

Two key approaches are needed to solve those problems. The first is to develop next-generation visualization tools that enable plant biologists to envisage and explore the complex operating networks and their outputs from database information. The second is to develop fully parameterized scalable mathematical models (both *deterministic* and *stochastic*) that describe every relevant biochemical reaction (transcriptional, protein dynamic, metabolic), cellular morphogenesis event (division, expansion, cell-cell interactions, organogenesis), environmental response (biotic and abiotic), and community interaction (involving multiple individuals and species) that can then be validated against the growing large-scale data sets (for example, temporal, spatial, and perturbagen response) collected by the NPGI community. Network models are then refined in a hypothesis-driven, experimental manner, and iterative refinement between modeling and experimentation generate clusters of evolving computable plants that are genuinely **predictive** of system behavior under any given condition—**the iPlant.**

The iPlant models will be a community product. They will depend on a substantial cyberinfrastructure to standardize and curate experimental data from both large-scale projects and individual investigator-driven data. The latter is an absolutely critical component, as model refinement and testing largely depends on the particular biological domain expertise of individual scientists and their students and colleagues. The models need to provide useful and detailed predictions to agriculturally relevant specialists, especially physiologists and breeders, in order to ultimately serve the economic interests of U.S. agriculture.

The iPlant will be constantly refined as predictions from models are tested, and will become the basis for creating highly effective novel plant strains for food, fuel, and fiber. This provides a detailed rationale for integrating germplasm, natural variation, and transgenes into the development and deployment of highly modified, or entirely new, chromosomes—the effects of which on the individual plant and its environment can be accurately predicted. Importantly, the iPlant will represent a collection of powerful research tools, allowing not only investigators to collaborate and explore extant data in an unprecedented manner, but also to help focus investment in experimental resources to the most productive lines of inquiry for both large-scale collaborative and individual investigator projects.

The iPlant will likely reveal biological principles that might also be found in similar endeavors in microorganisms and animals. Such common themes of biological systems architecture will accelerate biologists' understanding of all organisms, and shed light on how the knowledge can be used for the benefit of humankind. In addition, those common themes of modular biology will themselves help reduce the considerable computational power required for the success of a biological "moonshot" such as iPlant, as researchers can make simplifications and abstractions in the absence of fully parameterized equations for a given process. The iPlant will also present novel educational paradigms, as it can be used to generate many "virtual plants" that are far smaller algorithmic clones of the iPlant, to be used for effective education and community outreach programs not only for plant science, but for biological systems in general.

Goals for Research in Molecular Regulatory Networks (Recommendation 4)

5-year goals

- Gather homogeneous data sets, and develop dynamic gene expression databases, for one or two "core omics" models—with a focus on key, major developmental transitions (for example, seedling emergence, transition to and time course through, flowering) and environmental interactions (pathogen infection, a very few abiotic stress conditions).
- Collect temporal data for the above.

- Develop a plant MOD-ENCODE project.
    - Spatial transcription factor expression databases (laser captured dissecting microscope; single cell / FACS collected).
    - Binding site definition for all transcription factors.
    - Chromatin immunoprecipitation sequence of transcription factor targets (selected tissues, cells, and time domains).
    - Network analysis of gene expression patterns versus factor binding sites.
- Begin to create the protein interactome map for several key organs at discrete times in development. The following would need to be achieved in order to do this:
    - Develop a fluorescent tagged protein resource for each cDNA.
    - Create interactome over time map.
    - Adapt automated green fluorescent protein (GFP) imaging platforms to measure more plant tissues.
    - Create searchable database of community GFP imaging experiments.
    - Incorporate image analysis expertise from disperse fields (physics, defense, neuroscience).
    - Determine the number of unique cell types in one or two models. Curate collection of GFP enhancer trap lines to completely represent all cell types, with markers for nucleus, cytoplasm and plasma membrane in each plant cell type.
    - Initiate dynamic four-dimensional imaging database for cellular morphology through most developmental stages using GFP collections.
- Begin mapping posttranslational modifications.
    - Refine technologies for mass spectrometry-based phosphomapping in plants.
    - Expand phosphome for soluble and membrane proteins.
    - Match kinases with phosphosites using orthologous approaches.
    - Adapt enzyme activity profiling to measure enzyme activation (primarily kinases and proteases) under many growth conditions.
- Further analyze the ionome and metabolome (including hormonome) at discrete developmental time points.
    - Greatly expand application of technology platforms throughout plant science.
    - Perform spatiotemporal mapping of major ion constituents.
    - Define plant metabolic networks in key areas such as cell wall synthesis.

10-year goals

Phenomics

- Develop phenotypic response data to extend iPlant model data to a second tier of models to enable emergent, organismal properties to be incorporated.
    - o Develop assays for measuring real-time genetic and epigenetic changes under stress conditions, pathogen infection, other dynamic conditions.
- Continue progress in cyberinfrastructure and informatics.
    - o Create advanced, integrative portals to combine data sets and to make valid comparisons between distinct omics data sets.
    - o Create and make available intuitive tools for network reconstructions.
    - o Generate visualization platforms for plant data from molecules to ecosystems.
    - o Initiate iPlant project in selected signaling networks and organogenesis scenarios.

Structural genomics

- Generate 3D structure for a member of each unique protein family in plants.

20-year achievements

- Primary biochemical functions for all plant proteins will be known.
- Posttranslational modification of all plant proteins will be measured under many conditions.
- Plant interactomes will be defined in vivo, along with subcellular localizations.
- Chemical probes will be commonplace for many plant enzymes (inhibitors, antagonists, allosteric activators, agonists).
- Metabolic fluxes and ion levels will be known for all tissues throughout development.
- Regulatory networks driving morphogenesis and environmental response will be mostly known.
- iPlant interface will be fully operational, data collection and refinement will continue.
- iPlant algorithms will be developed and their accuracy and value will have been critically tested in many partial omics model species that represent major crop types and scientifically important species.
- Synthetic biology will be in widespread use in plant science.
- Remote sensing phenomics will be established for majority of world's crops,

reference plant experimental sites, and wide range of selected niches—automated and satellite controlled.

Evolutionary, Ecological and Communities Genomics

RECOMMENDATION 5: Increase the understanding of plant evolution, domestication, and performance in various ecological settings via investment in comparative genomics, and in the metagenomics of living communities of interacting organisms.

Selection of plants with particular traits by humans, the process of domestication, has increased production in crop species, altered many plant developmental characteristics in relation to the wild ancestors of crop species, and allowed humans to sculpt ecosystems dramatically through agriculture and urbanization. The selection process can be viewed as rapid, facilitated evolution for human use. A better understanding of the basis of phenotypic diversity in developmental, environmental, and evolutionary contexts will allow better manipulation of crop systems for improved agricultural productivity and enable the preservation of near-wild ecosystems. Some plant genome sequences are already completed and more are expected to be complete in the near future (see Chapter 2). In most systems, high-throughput approaches have been undertaken with the goal to analyze gene expression data for various tissues, organs, and growth conditions. The next challenge is to understand how the network of gene regulation gives rise to phenotypes (see Recommendation 3). Understanding the networks that ultimately determine plant form and function is greatly facilitated by comparison of natural, large-scale genomic sequence variation and gene expression data both within and across species.

Understanding and managing natural genetic variation is central to all efforts in crop improvement. Natural variation provides the basis for hybrid vigor, local adaptation, and biodiversity. In agricultural and natural populations, trait-genotype association studies and quantitative trait locus mapping will play a major role in analysis of trait variation. Advances over the next decade will revolutionize the understanding of trait variation in plants. Genomic analyses of genetic diversity are needed for association studies to discover agriculturally important genes and to serve as a source of genetic polymorphisms for functional analyses of plant biology.

Comparative genomics data cannot yet be fully exploited in plants, as has been done so successfully in the primate lineage for example, because of insufficient genome sequence coverage both within key model species and across

species from different parts of the green tree of life. Hence, the ability to understand how the evolution of natural variation sculpts plant form and function at a mechanistic level, and how that variation allows plants to exploit particular ecosystem niches at a population level, is currently limited largely by the amount of genome sequence available (see above).

Environmental Benefits of Plant Genomics Research

Photosynthetic organisms play a central role in all of the Earth's major ecosystems. As a result, understanding how plants function, and how to modify and improve their ability to carry out specific physiological processes—the ultimate goals of plant genome research—could have large benefits for terrestrial and linked aquatic ecosystems. Because NPGI programs improve the basic knowledge and infrastructure for understanding, managing, and breeding all plants, its environmental impacts will be deep and far reaching.

Genomics research is the fundamental enabling technology that allows plant scientists to optimize plant yields and quality as raw materials for energy production. Plants are a main converter of carbon dioxide in the atmosphere to fixed carbon. Engineering crops, including forest trees, to better sequester carbon offers the potential to increase the conversion of atmospheric CO_2 into soil biomass, ameliorating or even reversing the increase in atmospheric CO_2 generated by combustion of fossil fuels. In addition, there will likely be a dramatically expanded role for plants in the generation of liquid fuel alternatives to oil and its refined products. Ethanol from corn and sugar cane, though only a first-generation approach to alternative fuel with limited environmental benefits and supply capacity, has nonetheless validated the concept of using plant-derived biomass as a source of fuel for transportation. In the future, this is likely to come increasingly from lignocellulosic sources including grasses such as *Miscanthus* and switchgrass, woody tissues from fast-growing tree species, and cellulosic crop waste (DOE 2005a). These developments will not only slow the rate of depletion of fossil fuels and decrease the net addition of CO_2 into the atmosphere, but also provide a scientifically feasible, economically realistic medium-term solution for U.S. energy security and independence from foreign oil.

The first generation of genetically engineered crops, though not resulting from modern genomics research and NPGI *per se*, nonetheless relied on earlier genomic information, including GenBank and other electronic databases of gene function and conservation, similar to the myriad informatic resources that have since been expanded as a result of NPGI and related efforts. These earlier resources informed gene isolation, modification, and analysis. Experience from the first generation of these crops (summarized by Traxler 2004; Delmer 2005; Fernandez-Cornejo and

Caswell 2006) shows how genomic resources generated by NPGI can lead to substantial environmental benefits for agricultural systems. For instance,

- Varieties expressing insecticidal proteins from *Bacillus thuringiensis* often require fewer insecticide applications than nonengineered varieties, leading to improvements in the biological diversity within agroecosystems.
- Reductions in pesticide application, or improvements in ecotoxicological properties of pesticides, can reduce ecological impacts on farm workers and water systems (FAOSTAT 2007; Huang et al. 2005; Huang et al. 2002; Qaim and Zilberman 2003).
- Reductions in tillage associated with herbicide-tolerant crops appear to promote a number of environmental values, most notably improved soil carbon, reduced erosion, reduced energy expenditure from tillage, and improved wildlife habitat.

Because of the complexities of agroecosystems and human behavior, the overall environmental impacts from changes in plant varieties are difficult to predict. Nonetheless, new types of plants whose improvement is a direct result of genomics research either via conventional breeding or genetic engineering will likely provide substantial environmental benefits. Examples of benefits are as follows:

- Increased efficiency of extracting nitrogen, phosphorus, and potassium from the soil, or improvement in its availability to animals (Raboy 2007) would decrease use of fertilizer and lead directly to a decrease in energy use and in eutrophication of wetland ecosystems that drain agricultural lands, and associated lower costs of production.
- Increased drought tolerance and decreased use of water (Tuberosa and Salvi 2006) would reduce energy costs from irrigation systems, unsustainable changes to water tables, and salinization and associated dysgenic changes to soil quality in arid agricultural systems.
- Modifying lignocellulose chemistry in trees and grasses would lead to a diminished need for chemical or physical pretreatment during pulping or fermentation to produce ethanol, with an associated decrease in energy use, toxicity of effluents, and processing costs (Chapple et al. 2007).
- Ability to remediate toxic chemicals in the environment would improve either by sequestering them from the soil for harvest and removal, or by enzymatically breaking them down into less persistent or toxic forms (Doty et al. 2000).

Genomics and the Major Transitions in Plant Evolution

There are more than 250,000 species of plants, representing a wide variety of growth habits, adaptive responses, and potentially useful traits. Modern plant genetics and genomics can provide experimental and computational tools for comparing genomes and for providing insights about the similarities and differences among organisms, the basis of ecologic adaptations, and their origins and persistence. There is untapped value in natural variation as a source of functional information since variant alleles can express altered function that might not be phenotypically uncovered in typical forward or reverse mutant genetic screens because of functional redundancy or epistasis. Furthermore, natural variants are enriched for alleles that exhibit high fitness in harsh natural environments, including associations with microbial communities that can influence plant productivity.

Many important issues in evolutionary and ecologic genomics can be addressed through comparisons among a diverse array of sequenced species. In the 2002 NRC report, the development of genomics resources in species outside the model species *Arabidopsis*, other reference species, and their crop relatives was advocated. In response to this recommendation, NPGI, largely under the auspices of JGI and following JGI's community-based, peer-reviewed system, undertook sequencing of reference genotypes of the plant species listed in Chapter 2. For several of these species, NPGI is positioned to take advantage of large collections of natural variation in the form of diverse cultivars, accessions, and wild relatives within each species and next-generation sequencing technologies. The natural variation available encapsulates various important plant characteristics including, among others, perennial and annual life cycles, broad differences in the ability to grow in different ecosystems and respond to pathogen and abiotic stress, different mating systems, and different ploidy levels.

Research in plant evolutionary genomics can build on resources developed in NPGI and the various *Arabidopsis* programs and has great potential to bring fundamental scientific advances in years to come. Maintaining a balance between crop-centered and nondomesticated plant species and populations is critical because diverse experimental systems expand scientists' ability to examine hypotheses and biological functions. In addition, genomic analyses of the wild relatives of crop plants could identify agriculturally important sources of genetic diversity. Those analyses are essential to understanding the responses of forests and grasslands to changes in climate and other environmental conditions. Both intra- and interspecific comparative genomics approaches will allow unique analyses of questions in plant biology, which would be difficult or impossible outside of NPGI because of associated genomics costs.

To ensure that the genomics investment in additional species builds effectively on existing resources, it was recommended (NRC 2002) that the evolutionary-genomics community pursue the selection of a modest number (10–20) of key species spanning critical evolutionary nodes in preparation for community-wide genomic investigation over the next 10–15 years. In response, several large community-driven projects concerning the evolutionary and ecological genomics of a cluster of crucifer species surrounding *Arabidopsis thaliana*, including *A. lyrata*, *Capsella*, *Brassica*, *Thellungiella*, and *Boechera*, have recently begun. These important genome sequences will significantly enhance the ability to define gene function and genome evolution trends, and will significantly expand the utility of *Arabidopsis* as a broadly relevant model. As well, the sequencing of two additional species, *Mimulus* and *Aquilegia* (Columbine), will open new genomics-based vistas in systems that are historic and well-characterized models of evolutionary and ecological plant biology.

In addition, the genome sequencing of a moss, *Physcomitrella patens* (a good experimental model for an important lineage of plants), and a lower vascular plant, *Selaginella*, fill important gaps in our understanding of plant evolution and diversity. Further effort to generate bacterial artificial chromosome (BAC) libraries from 22 nodal species and 12 wild rice relatives were also funded (NSF 2007c). This trend is laudable and should continue because it provides an excellent way to leverage reference genome sequences and focus efforts using reference species to determine gene function and network operation to maximum effect. For example, the imminent completion of the first reference maize genome would open the door to sequencing of close relatives such as *Zea diploperennis* and *Zea luxurians*. The diploid *Tripsacum dactyloides* would be valuable to validate gene predictions and regulatory sites in maize and to understand evolutionary trends over a 4–5 million year evolutionary timeframe.

The overall goals of evolutionary and ecological genomics, as articulated above, have been readily adopted by the *Arabidopsis*, rice, and maize genetics communities, where a history of community-based resource development has been facilitated by NPGI and the NSF's *Arabidopsis* Genome and 2010 Programs. Consensus building within the evolutionary biology community led to community-developed goals and subsequent sequencing efforts at JGI on the species described above. The potential value of these exercises to crop improvement cannot be overstated: The more that is known about how the diversity of plant life evolved, the better equipped scientists will be to manipulate basic plant processes for practical and sustainable benefit. These facts provide additional support for the argument that an expanded role for JGI, or another, equivalent large scale sequencing center, is critical to the overall success of NPGI.

While efforts to date are laudable, the committee strongly recommends a significant expansion of evolutionary and ecological genomics, and of the study of plants in association with other organisms (interspecies interactions genom-

ics). Tools for genomic studies of diverse evolutionary and ecological focal species need to be explicitly comparative and developed with nonspecialists in mind. The aim would be to broaden the community of researchers who have access to and can effectively use plant-genomic tools and data in the future. Technologies such as low cost resequencing, initial reference genome sequencing, and association genetics greatly lower the threshold for which new species can be considered useful experimental organisms. The iterative reduction in costs is relevant for both plants and for the associated organisms that can add to or detract from plant performance. Hence, the set of desirable criteria for choosing new reference species for evolutionary, ecological, and "interspecies interactions" genomics programs would include (for *both* the plants and where relevant the "interacting organisms" under study):

- Distributed position in the phylogeny and clustered locations to bring added value to related, sequenced species.
- Genome size, with emphasis on small genomes and, at first, simple ploidy.
- Genetic tractability.
- Ease of crossing and population development.
- Ease of growth.
- Short generation time, except where this is precluded due to interest in ecologically or economically important species (for example, perennial or woody species).
- Availability of existing tools such as germplasm collections, mapping populations, genetic and physical maps, large insert clone collections, mutant collections, and genetic transformation.
- Size of the research community versus required investment.
- Economic importance of the focal species or close relatives.
- Exceptional properties of interest, such as selected species that have undergone recent polyploidization.

Goals for Plant Evolutionary, Ecological, and Communities Genomics (Recommendation 5)

5- to 10-year goals

- Sequence species representing the major phylogenetic nodes of plant diversity and clusters of relatives to bring added value to related, sequenced species.
- Sequence microbial species representing the major phylogenetic nodes of pathogen, commensal, and symbiotic diversity, and clusters of their relatives.
- Sequence and characterize several plant associated metagenomes from cul-

tivated and natural systems using well-understood model systems as departure points.

- Improve understanding of natural genetic variation at the levels of nucleotide polymorphism and trait variation in selected species.
- Develop representative models to address major evolutionary issues, such as speciation, adaptive trait evolution, adaptation to environmental conditions, and evolution between and within population.
- Study domestication in crop plants to understand evolutionary principles, agriculturally important genes, natural genetic variation among cultivars and landraces, adaptation to heterogeneous environments and changing climates, and phylogenetic differences among plant groups.

10-year goals

- Develop representative systems to analyze how plants interact with their associated microbial and insect communities (from pathogens and herbivores to commensals and pollinators), in the laboratory and in ecological settings, both natural and cultivated, over co-evolutionary time scales.
- Study in depth the developmental and adaptive mechanisms that are ecologically or economically important. Studies would include the diversity of adaptations to abiotic and biotic stresses, development of complex structures such as wood and fruits, and mechanisms for control of complex life cycles and phase changes.
- Explain core evolutionary principles, such as speciation and adaptation mechanisms, the role of genetic drift and of reproductive systems on evolution, and the evolutionary processes that influence complex trait variation.
- Develop new informatics tools for massive data sets and for integration of diverse data types, including genomes, pathways, phenotypes, and environmental context.
- Develop efficient transgenic methods to bring to nonmodel species.

20-year achievements

- Pan-genomic data—that is visually intuitive, manipulable, useful, and quantitative—will be integrated and easily accessible.
- The role of natural and artificial selection in shaping genetic variation and genome structure will be understood.
- The role of epigenetics in development, heterosis, adaptation, evolution, and genome architecture will be explained.
- The mechanisms and consequences of polyploidy and copy number variation will be explained.

• The convergence of systems biology and evolutionary biology will be facilitated by integrating ecologists with genetics, genomics, and evolutionary biology communities.

Translating Plant Genomics into Plant Improvement

RECOMMENDATION 6. Enable translation of basic plant genomics towards sustainable deliverables in the field, and continue to use NPGI as a foundation for new, agency-specific, mission-oriented plant improvement programs.

The first 10 years of the NPGI have made a strong start toward understanding the fundamental challenge of how plants work. In order to most effectively translate knowledge from the basic science at the core of NPGI into commercial innovation, and to accelerate the pace of translation to practical outcomes, additional enabling tools and methods for enhanced transfer from model systems to crop species should be developed. Given the considerable fundamental advances in understanding plant genomes and the substantial progress in translation (discussed above) since NPGI's recent addition of this area of research to its portfolio (NRC 2002), the arguments for a continuation and expansion of translational genomics are clear and strong.

A broad genomic survey—beginning with draft genome sequencing and expressed sequence tags—is needed in as many crop species as possible to enable translation. These genomic surveys can be prioritized on the basis of criteria including: taxonomic diversity, the likelihood of the information gathered being "pulled through" into commercial breeding by a critical-mass community of scientists (in both the private and public sector), and the economic footprint of the crop. The prioritization is analogous to the differentiation between investment in "full" or "partial" omics toolkits outlined above in Recommendation 2.

Challenges Translating from Models to Multiple Species with Specialized Traits

For many of the traits in high demand by consumers or important to industry, model systems provide only very basic leads. They include nutriceutical-associated traits such as antioxidants, fiber, vitamins, and allergens; consumer acceptance traits such as aroma, texture, and ripening; vegetative architecture, woody tissue structure, and growth habit; adaptations to specific and complex abiotic and biotic stresses; and crop-specific domestication, which might differ significantly from the domestication traits that have been studied in cereals.

Trees present special challenges for translational genomics. Commercially valuable tree phenotypes are typically polygenic. The unspecialized manner with which most wood and pulp feedstocks are grown in the United States reduces the value of genomics investment. The outcrossing and heterozygous structure of breeding populations—coupled with multiple-year delays to onset of flowering—makes genetic introgression difficult. Marker-assisted breeding might not be cost-effective even where strong marker-trait associations are discovered (Johnson et al. 2000; Strauss et al.1992). A large number of genotypes are typically used in production plantations. There are high costs in time and space to gather statistically reliable phenotypic information about growth, physiology, and wood properties. For these production systems, the costs and efficiencies of genotyping and phenotypic assessment methods have to continue to improve if translation is to be effective.

The Unique Challenges of Plant Breeding

Plant genomics has specific requirements that can differ significantly from the requirements of human genomics, for example. More importantly, marker-assisted breeding projects benefit from large plant population sizes (tens of thousands to millions)—the more individual recombinants that can be tested in an introgression program, the fewer generations are required to cleanly recombine the genes of interest from two or more genomes. However, because marker-assisted breeding in plant biology must span many species (compared to human population genetics, where only one species is relevant), there will continue to be large costs associated with genomic sequencing and subsequent definition of a robust SNP set for association mapping. Additionally, it would be ultimately desirable to develop methods for SNP detection and mapping "in the field," and this is likely to drive technology development that differs from the technologies now being successfully applied in human genetics research.

Accelerating the Practical Application of Plant Genomics: Markers, Markers, and More Markers

Clearly, plant scientists' thirst for high-throughput, cost-effective DNA markers has not been quenched. Marker technology empowers gene and quantitative trait loci (QTL) discovery and marker-assisted selection in crop-breeding programs, and facilitates the study of plant evolution across all taxonomic levels. The application of marker technologies has significantly advanced the study of both single- and multi-genic traits of economic importance, has become a standard tool for many plant breeders, and has been widely employed in research studying the genomic conservation among related plant genomes and the evolution of plants. However, markers are not inexpensive enough to assay or abundant enough to be employed for breeding in most economic plant species. Likewise, the infrastructure, informatics, and expertise for most of those species are not adequate to maximize economic benefit from markers.

The decrease in cost of DNA sequence acquisition, which leads to an ever-increasing availability of DNA sequence for crop plants and their wild relatives, will provide a template for additional SNP discovery. In some of the major crop species, resequencing of multiple genotypes will enable massive SNP discovery, and will provide the first steps toward enabling association genetics in these important crops. SNP marker development could be expanded for a number of crop plants that are of considerable economic importance but do not occupy major U.S. land areas and for plants that are being targeted to meet the nation's emerging biofuel needs. Forest tree species, particularly the conifers with their exceedingly large genomes, will not be feasible to sequence completely with current technologies, but survey sequencing to identify nucleotide polymorphisms is already underway and further support is worthwhile and beneficial.

An essential component in turning the polymorphisms discovered by comparative sequencing into useful breeding tools is the development of detection technology that can be deployed in multiple species at low cost. Although polymorphism detection tools are well developed in the private sector, particularly for the large commodity crops, a major emphasis on developing and disseminating technology that can be cheaply deployed in relatively understudied species is necessary. Providing breeders with training in the applications of these newly developed detection technologies in ongoing cultivar development programs is critical.

Understanding Plant Phenotypes at Multiple Resolutions

Economic plants are grown for their unique and valuable phenotypes. Manifestation of those phenotypes is a result of the interaction between specific genotypes and their environment. The environment affects plant growth and development, the biochemical composition of plants, and can even result in a change of heritable genetic variation. Scientific understanding of plants grown in the context of their environment is inadequate at present. Crop models need to be studied in the diverse set of environments that they might experience in production settings to generate predictive models. Functional and comparative genomics offer powerful tools to understand the mechanisms controlling interactions of plants with their environment, and the underlying genes and gene interactions. However, without a high level of precision in the collection of phenotypic data, the ability to determine genotype will vastly outpace the ability to associate it with a specific phenotype. One component of high-resolution phenotyping will be the development of novel biomarkers that accurately report the physiological status of the plant in the context of its environment in real time and the biotic or abiotic stresses it encounters. As genomics research is moved to the field, expertise in supporting areas including informatics, statistics, and quantitative biology will be required. Research programs that integrate genomic scientists, plant breeders, ecologists, physiologists, informaticists, and engineers will be able to measure and interpret massive data sets that describe real-time growth and development of plants in the context of complex environmental signals and challenges. The programs need to include the development of novel engineering solutions for remote sensing of plant physiology from the population (field) level down to the level of an organ of an individual. Because production crops and systems are diverse, numerous plant model systems will be needed to fully explore the breadth of economically important plant systems. Examining the feasibility of creating centers where standardized, efficient, high-throughput phenotyping technologies can be applied to several major crop and tree species would be a logical next step. Such centers are already established in the private sector.

Producing Plants That Overcome the Limitations of Their Environment

NPGI targeted biotic stresses as part of its initial translational genomics effort. Initial successes in the understanding of plant responses to biotic stress demand that this area of research be scaled up and expanded to include abiotic stress improvement as a core goal for the next 10-year cycle of NPGI. The ability to sequence multiple isolates of a single plant pathogen species cost-effectively, coupled with the availability of dense marker maps in several crop species as well as large mapping of natural populations, allows detailed studies of the interaction of crop plants and their pathogens and their insect enemies. In addition, dissecting the genetics of traits that allow plants to perform in suboptimal environments (for example that are critically limited by water, temperature, nutrients, and light) is particularly important in preparation for a changing global environment. Developing innovative approaches to phenotype large plant populations with the necessary throughput, precision, and cost-effectiveness to provide replicated data collection within and across multiple environments is critical to studying plants in such environments.

Unlocking the Genic Treasures Within our Plant Germplasm Collections

The United States spends considerable resources in the collection, maintenance, and characterization of crop germplasm collections. These collections have been routinely used to identify and deploy the genetic diversity necessary for plant breeders to respond to pest adaptation, the unintended import of new pests, and changing consumer demands. The collections also contain many desirable alleles for more complex traits, such as seed protein content and productivity. Historically, the approach to finding desirable alleles was based solely on the accession's phenotype. A quantitative trait by its very nature is the result of the effects of multiple genes and likely their interactions.

One potential of densely saturated DNA marker maps in regions surrounding important QTL of crop species would be applications in identifying germplasm accessions with additional allelic diversity for those QTL, or the identification of accessions that likely possess positive alleles at previously undetected QTL. Research is encouraged into novel approaches to mine the U.S. germplasm collections for unique alleles and introgress them into improved cultivars. Hence, a better understanding of the wealth of genes present in these collections via genomic analyses and more effective methods for their deployment would be worthwhile.

Reaping Rewards from Genome Comparisons

Plant biology has a vast phylogenetic sweep. As such, genomics increasingly becomes a unifying theme for plant biology. Plant breeders can take advantage of the variation across the genomes of closely related populations, as well as the distant relatives of crop species. The availability of genomic resources and a clear understanding of the evolutionary divergence of various plant species can facilitate the identification of agriculturally important genes in crop species through the knowledge of its genomic location in a model species. More research in comparative genomics is necessary to expand the value and utility of the model plant species, as discussed under Recommendation 1, above. As a five-year goal, NPGI should provide tools for the successful deployment of improved crop cultivars from improved alleles discovered from comparative mapping.

Integrating Informatics Resources for Breeders, Evolutionary Biologists, and Ecologists

The importance of informatics cannot be overstated in any genomics program. Informatics will become increasingly important as plant biologists attempt to push the fundamental findings of NPGI toward applications that benefit human welfare. The last five years have seen advances and even mergers in the development of plant sequence-based and breeder-based databases. Although there is much more to achieve, the recent advances have been empowering to plant scientists. The development and implementation of software programs that clarify the relatedness of individuals within a study population and enable identification of associations between DNA markers and phenotypic traits have advanced efforts to find the genes underlying important phenotypes in the U.S. crop germplasm collections. In addition, these research advancements are now being applied to study the genetic makeup of natural populations in the environment.

Plant genomic databases have become standard tools of genomic researchers, and the advances in translational genomics have enhanced their utility to crop breeders and environmental scientists. On one hand, the ability to compare DNA sequences across plant species is essential to leverage information from one species to another. On the other hand, plant breeders need to be able to navigate from sequences to genes to phenotypes within a single species. Although many plant genomic databases contain similar sorts of information, the way this information is accessed by genomicists and breeders is often different. To address the needs of diverse users, the most effective and efficient approach might be for database developers to create different entry portals for a unified database. Such entry portals would provide scientists of various backgrounds a user-friendly interface that enables direct access to relevant data (for example, multiple ways to access and manipulate the data well). For crop breeders, the ability to navigate from maps to genes to traits to plant accession by augmenting existing databases with phenotypic data (to include expression data QTL) and genotypic data is important.

Enabling Plant Improvement Through Plant Transformation

NPGI has not made a significant investment in the development of transgenic tools to support functional and translational genomics. Many of the advances that will soon be possible as a result of the rapidly advancing knowledge of plant genomics will require the ability to routinely, cost-effectively insert modified genes, or replace alleles, in a number of different plant species for research or for deployment in breeding populations. Effective transgenic tools will, at the minimum, be required to rigorously test the function of genes identified via mapping or association studies. Beyond basic descriptions of gene expression, transgenic methods are generally the most powerful means for determination of gene function due to the wide variety of modifications to expression that can be introduced and studied. The practical exploitation of many of the advances made possible by genomics will be best carried out by introducing a desired trait in a highly specific way without the limitations of generation time or the feasible size of breeding populations. Transformation is especially valuable in outcrossing and long-lived crops with complex genomes, such as trees and grasses important for bioenergy, because of their long generation times and intolerance of inbreeding.

Transformation does not need to be used solely for wide transfers of genetic information, which are the applications that elicit the most social concern. It can also be used to modify native genes or patterns of gene expression, and to increase trait diversity in highly bred or elite clonal varieties. Certain technologies—such as "all-native transformation" (Rommens 2007) that relies on plant-derived promoters, plant genes conferring traits of interest, marker-free integration, and "P-DNA" sequences sufficiently similar to T-DNA borders to mediate DNA delivery—are among the relevant new transformation technologies that may find higher levels of social acceptance among the wider public, and especially the organic and sustainable agriculture communities that have mostly rejected transgenic technologies to date.

Artificial Chromosomes: Stacked Traits in a Breeding Block

Current transgenic techniques are ill-suited to the engineering of complex pathways and multigenic traits because they can introduce only few genes at a time. Moreover, these genes are inserted into the host chromosomes creating unpredictable effects for both host gene function and the function of the transgene. Plant artificial chromosomes (or minichromosomes) offer the potential to engineer entire pathways and stack independent traits efficiently in one, easily controlled autonomous DNA molecule that does not integrate into the host chromosomes. Similar technologies in yeast and bacteria (yeast artificial chromosomes and BACs) were revolutionary. Engineered minichromosomes, made through de novo artificial chromosome construction or by natural formation (Yu et al. 2007), will provide a similar quantum leap forward for plant engineering. They will enable the delivery of gene stacks to improve biofuel feedstocks; to produce food crops that are resistant to multiple diseases, pests, and abiotic stress; and to generate plants capable of synthesizing molecules useful for medicinal or pharmaceutical applications. Realizing these potentials depends on developing minichromosomes in multiple plant species that can be easily engineered and function with high fidelity. That development will require a combination of practical plant biotechnology and basic research into the mechanisms of chromosome structure and function.

Goals for Translation (Recommendation 6)

5-year goals

- Broaden genomic DNA sequencing survey to as many crop species as possible.
- Expand marker development, SNP marker discovery, and develop polymorphism tools that can be deployed in multiple species at low cost.
- Develop novel approaches to mine unique alleles from existing germplasm.

• Deploy improved crop cultivars from improved alleles that were discovered as the result of comparative mapping between model organisms and crops.

• Test new approaches to improve transformation and allele-replacement capabilities.

• Amass and analyze temporal and developmental catalogues of mRNA and small RNA populations under different growth and stress conditions.

• Conduct feasibility studies on the creation of a few of centers where standardized, efficient, high-throughput phenotyping technologies can be applied to several species.

10-year goals

• Develop novel biomarkers that accurately report the physiological status of the plant in the context of its environment in real time.

• Develop crop plant genomics databases that easily access complete genotypic and phenotypic data in the field.

• Establish public field evaluation centers to integrate these data from biochemistry to agronomy.

• Create data-mining and modeling approaches for dissecting complex trait architectures, and visualization tools for understanding these architectures.

• Create rapid field-based phenotyping approaches and field-based physiological sensors that include remote sensing of individual plant performance.

• Establish models that accurately predict agronomic phenotype from genotype.

• Demonstrate efficient site-directed transformation and allele-replacement capabilities in a limited number of species

• Develop artificial and replacement chromosome technology for major crop species using alleles of specific genes for specific traits.

• Develop containment mechanisms for transgenes and artificial chromosomes that function across a range of species.

20-year achievements

• Accessible, useful, integrated informatics resources and tools will be available and used by breeders, biotechnologists, and environmental scientists.

• Artificial chromosomes and transformation will be used routinely as a precise mutagenesis and gene transfer tool, in combination with a customized suite of containment methods when needed, in a wide variety of crop species.

• Plant biologists will have the capability to produce a plant with specific performance characteristics.

- Synthetic research and crop plants will be widely disseminated.
- Land and air based remote sensing of biomarkers (phenomics) will be deployed for the majority of world's key crops, at experimental sites and wide range of selected niches—fully automated and satellite controlled.

Informatics, the Mesh That Envelopes Plant Genomics

RECOMMENDATION 7: Develop and deploy sustainable, adaptable, interoperable, accessible, and evolvable computational tools to support and enhance Recommendations 1–6.

NPGI's future success relies on the development of the computational methods, tools, and databases that enable the integration of disparate data from multiple technology platforms and geographically distributed laboratories. These methods would provide a wide variety of means for collecting and collating data, create new visualization paradigms, and enable researchers to quickly integrate new analytical methods. The successful implementation of computational infrastructure will ensure that every genomic effort has useful payoffs that are beyond the original goals of individual projects by providing an overarching structure to access, manipulate, and view the data flowing out of each endeavor.

Multiple perspectives are necessary to generate a complete picture of the informatics needs. This section is divided into three views on the challenges: data collection and exchange, data management, and organizational requirements that provide a means of involving the plant genome community. The common thematic element of the three views is the importance of open standards and specifications that are widely published and are developed through community efforts. Examples of current standardization efforts that should be supported and emulated are the Gene Ontology, the Plant Associated Microbes Gene Ontology (PAMGO, a domain-specific contributor to the Gene Ontology), the Open Biological Ontologies, and Plant Ontology efforts. The integration of knowledge and the interoperability of tools are only possible through a sustained commitment to standardization. Standardization ensures the reproducibility of computational analyses, and a high degree of confidence is critical for the further accrual of knowledge. It also reduces the cost and risks to plant genome research by removing the reliance on individual developers and suppliers, providing more flexibility in software development, enhancing the integration of systems, and investing the ownership of the informatics in individual members of the community. One role of NPGI will be to support standardization efforts because they benefit the entire plant research community.

Data Collection and Exchange

Pattern recognition, from which inferences are drawn and hypotheses are formed for subsequent experimental testing, is best achieved from uniformly gathered, high-quality data. Improved informatics technology does not necessarily alter the process by which the information is gathered, but it improves the speed and reliability of inference generation from larger and larger volumes of data. Data typically flow from their original collection and processing to an organized repository where they can be retrieved and analyzed. Hence, technological improvements can enable scientists to find important connections between data that previously were maintained in isolation.

Collection

Broadly speaking, the resolution of the information collected correlates with the experimental approach; observation of individual samples provides highly nuanced and detailed information over a limited number of cases, whereas high-throughput studies in which multiple samples are examined concurrently cover a larger portion of the genome but with a certain loss of detail. NPGI aims to integrate these strategies so that research data are combined to generate highly detailed information across the breadth of entire genomes. The term "well-annotated" infers three characteristics. First, the annotation comprehensively covers the entire range of genomic features from protein-coding genes to functional RNAs to regulatory regions to binding sites. Second, the annotation is complete: That is, every occurrence of every different feature type is identified and precisely and accurately located. Third, all of the ancillary biological data associated with these features (such as phenotypic descriptions, time of expression, location of expression, and so forth) are richly described. Because the genome sequence provides the backbone resource on which all other resources depend, its utility depends on it being as comprehensive, complete, precise, and detailed as possible.

Ideally these data should be captured at the source, including as much biological context as possible and integration with existing data sets in real time; delay often leads to omissions and inaccuracies. Portable genotyping devices are on the foreseeable horizon. These devices will incorporate cameras, image recognition software that allow the specimen to be compared to similar plants, capability to access locally installed and remote databases, and global positioning systems so that the location of a sample is automatically collected. Tools and standards need to be developed that allow pan-genomic data to be collected in such a way to enable data integration that is accessible—visual, easily manipulated, and quantitative.

Exchange

There are many technical challenges to data exchange. For certain key components, after developing and testing alternative methods, there was adoption of an effective community standard that provided the solution. Widely distributed data are only accessible if there are transport mechanisms in place that use standard protocols for retrieving information. The success of the Internet is based upon the universal acceptance of a single transport mechanism as the standard—the transmission control protocol and Internet protocol. Similarly, standardization upon common query languages such as structured query language has made efficient query of immense datasets possible. Convergence on shared data compression techniques has provided the speed and performance needed to efficiently retrieve voluminous data sets from remote locations. Common syntaxes, such as extensible markup language, allow basic syntactic parsing of the data that is collected. The outstanding challenge is to interpret the data semantically, which requires the development of a shared descriptive language to enable the data integration needed for plant genome research.

All stakeholders need to be involved in the development of these standards. Standards cannot be dictated by the developers, but they have to evolve based on real-world usage. In their development, the standards need to be applied to actual data sets being generated by researchers.

Data Management

Databases permeate the plant genome community. They have three important roles: for information management and analyses, as community resources, and as archives. Databases are invaluable for laboratory information management and experimental analysis. High-throughput experiments and geographically distributed projects would be difficult without a database component for managing the data that is generated. Databases also provide a centralized community resource (for example, TAIR or Gramene databases). Laboratory databases need to have a clearly articulated description of how and when the transfer of data into the community repositories will be achieved. Archival databases are distinguished from "in use" databases in that the latter use the most accurate information currently available to draw inferences for building experimental hypotheses. Archival databases typically include all the data as a historical repository to understand the path to the current best available data. Hence, the most useful community databases, such as TAIR and Gramene, clearly indicate how they distinguish between archival and current data.

Organizational Requirements and Involving the Community

To ensure adoption, it is essential that plant biologists direct the building of the bioinformatics resources, tools, and standards. Without such involvement even the most computationally sophisticated tools will not likely be used. Successful implementation of informatics technology for plant genomics would require the technology to be:

- Sustainable. A mechanism for maintaining a useful knowledge environment over time is necessary.
- Adaptable. The informatics resource needs to continuously absorb and incorporate new knowledge in its subject domain.
- Interoperable. The components of the system have to be easily integrated with each other, so that new knowledge can be gained by comparative studies across multiple data forms and levels of biological organization.
- Accessible. Information and annotations need to be interpretable by a wide variety of plant scientists, breeders, and students with basic informatics training, not just genomic scientists.
- Evolvable. The informatics resource needs to quickly adapt to the requirements of the plant genomics research community and these requirements would be the prime selection pressure on the technology.

The large amount of information housed in genomics databases and the expected explosion in data generate pressures on the organization of research to effectively mine that data. The plant research community will have to place greater emphasis on integrating bioinformatics approaches into its work. The committee proposes a national strategy for bioinformatics that includes training, collaboration with large data centers, and bioinformatics-oriented research, such as the creation of specialized databases, new analytic tools, and semantic standards.

Goals for Informatics (Recommendation 7)

5-year goals

- Develop and adopt open semantic standards and data exchange specifications for the plant genomics community.
- Develop software utilities for using these standards to describe data as they are collected initially in individual laboratories, including the use of widely understood terms and language to promote accessibility.
- Develop an architectural infrastructure so that each project could explain how it fits into that architecture.

- Develop new informatics tools for the integration of massive data sets of diverse data types, including genomes, pathways, phenotypes, physiology, and environmental context.

20-year achievements

- Plant biologists will have comprehensive knowledge environment for plant genomics that comprises highly detailed biological information across the breadth of entire genomes.
- Data-capturing tools that are fully integrated with the knowledge generation by the researchers who collect the original data will be available; the tools will have a seamless operation from data generation to data management.

EDUCATION AND OUTREACH

Education

RECOMMENDATION 8: Improve the recruitment of the best broadly trained scientists into plant sciences.

In the last decade, public awareness of the dramatic advances in biology has focused primarily on medical applications and controversies. The high profile of the Human Genome Project and explorations of medically relevant questions have overshadowed progress in plant genomics, at least in the eye of the public. Attracting new and diverse scientists to plant genomics will likely necessitate reaching out to students who might not have considered plant biology or genomics, or indeed biology at all, when they commenced their training.

Converging challenges—environmental degradation and resource limitations, climate change, increasing demand for food, and demand for renewable energy sources—will require solutions that are based both in conservation and technology. The latter will require increased numbers of appropriately trained scientists over the next 20 years. While all STEM disciplines will provide some of the solutions to these global problems, plant science has a leading role to play. A well-known catalyst to attract highly skilled scientists to important and emerging research areas is funding, particularly at early career points. Graduate-level and postdoctoral fellowships are among the best mechanisms to recruit talented scientists into a field. It follows, therefore, that targeted Ph.D. and postdoctoral fellowship programs, within the context of NPGI, will infuse the plant science community with more and better young scientists to address key societal problems.

Given that plant genomic analysis is becoming increasingly interdisciplinary, new scientists should be given incentives to enter the field from computational disciplines, such as computer science or statistics. For example, as genome sequencing expands to capture diversity and natural selection, more students with backgrounds in ecology may also potentially be drawn to plant genomics. International cooperation will likely continue to play a significant role in translational genomics, given the expanding repertoire of plant species for which genome sequence and genomic tools are available, coupled with the increasingly global nature of scientific inquiry, agricultural markets, and global concerns about sustainability.

Despite the large number and expansive scope of education and outreach initiatives, metrics to measure their impact and success against their goals are essentially lacking. Although there is a pervasive expectation among reviewers and panelists that grant recipients would engage in some form of outreach activity, there is little expectation that the efficacy of that outreach will be formally assessed. Developing metrics to assess outreach initiatives would help funding agencies evaluate whether the investments were well-made. **It is essential that funding agencies institute rigorous reporting requirements, mechanisms to track the longer-term career paths of trainees, and standardized metrics to assess outreach initiatives, so that they can evaluate whether investments were well made.**

NPGI should be a leader in education of interdisciplinary scientists. The recommended future directions for plant genomics, outlined above, will require a cadre of students who bring broad skills and knowledge to a wide-ranging set of problems. In particular, the next generation of plant genomics practitioners needs to be adept at computational and systems-levels approaches, and at least comfortable with modern scripting languages. As the nature of research and the questions posed throughout the discipline shift dramatically to include data sets requiring sophisticated statistical analysis, future students will need the confidence to approach biological phenomena quantitatively (Bialek and Botstein 2004).

At the graduate and postdoctoral levels, the need for computational expertise currently often entails *ad hoc* arrangements to bring students trained in one of the disciplines "up to speed" in the other on a need-to-know basis, although methodological and cultural differences between the fields of biology and computer science pose significant challenges (Zauhar 2001). A variety of curricular models have evolved to bridge gaps in communication between biologists and computer scientists (Dyer and LeBlanc 2002; Gerstein et al. 2007; Pevzner 2004), and there has been rapid growth in the number of graduate and even undergraduate-level programs in computational biology and bioinformatics (Zauhar 2001). Centers and departments devoted to systems biology represent perhaps the most high-profile approach to integrating computer modeling, large-scale data analysis, and empirical biological research (Check 2003; Ideker 2004). At least in the short term, however, those programs are unlikely to meet the dramatically rising demand for scientists with interdisciplinary expertise that includes bioinformatics (Zauhar 2001).

The NRC report *Rising Above the Gathering Storm* (NRC 2007b) recommended increasing the number of U.S. citizens pursuing graduate study in "areas of national need" by funding 5,000 new graduate fellowships each year. NPGI should build mechanisms to ensure that the number of graduate and undergraduate students with rigorous training in both biological and quantitative approaches to plant genomics is sufficient to support a thriving research and development job environment in both the public and private sectors. By leading with new opportunities for graduate support in bioinformatics and computational biology within the context of plant genomics, NPGI could bolster the image of plant science as an exciting alternative to the biomedical fields for ambitious and creative students.

Students trained in engineering and computational sciences might represent an untapped resource whose skills and inclinations could make them valuable contributors to plant genomics. In particular, engineers are familiar with systems that behave imperfectly, and their systems-level perspective has already enabled important strides in modeling biological regulatory circuitry (Wiley et al. 2003). Collaborative relationships with faculty in engineering could lead to unique training opportunities. NPGI researchers interested in establishing connections with colleagues in engineering might consider the emerging field of synthetic biology (Endy 2005) as a possible example of common ground.

As a growing number of Ph.D. umbrella programs require a course in genomics and bioinformatics of all their students, it is incumbent upon NPGI-funded faculty members to insist on standards in their institutions' training programs that will meet their research needs. NPGI-funded PIs could be encouraged to offer modules or other shared teaching formats in these courses.

Eventually, most incoming graduate students will be able to fulfill requirements in bioinformatics through their undergraduate education, and important inroads have already been made in developing bioinformatics curricular materials suitable for even introductory-level biology courses (Honts 2003; Campbell and Heyer 2007). However, until those fields trickle down to become standard course offerings at the undergraduate level, graduate programs will need to provide them to the incoming students. Summer internships are one path by which interested undergraduates can become acquainted with, and gain proficiency, in bioinformatics and computational biology. Table 3-2 lists several such programs.

The NRC report *BIO 2010* advocates encouraging all students to pursue independent research as early as possible in their career (NRC 2003). These research experiences reinforce, clarify, or increase students' interest in postgraduate education (Lopatto 2004; Seymour et al. 2004) and can result in enhanced confidence in

TABLE 3-2 Representative REU Programs and Summer Internships for Undergraduates in Computational Biology, Bioinformatics, and Systems Biology

Field	Institution	Website
Computational and systems biology	Iowa State University	*http://www.bioinformatics.iastate.edu/BBSI*
Computational and systems biology	Massachusetts Institute of Technology	*http://csbi.mit.edu/website/outreach_programs/summerintern*
Computational biology	University of Connecticut Health Center	*http://www.nrcam.ucnh.edu/news/positions.html#intern*
Bioinformatics/computational biology	University of Maryland Baltimore County	*http://www.umbc.edu/SPCB/*
Bioinformatics/genome science	University of Southern California	*http://cegs.cmb.usc.edu/academics/bigs/BIGS.html*
Computational biology/bioengineering/Bioinformatics	University of Pittsburgh	*http://www.ccbb.pitt.edu/BBSI/index.htm*
Bioinformatics/bioengineering	Virginia Commonwealth University	*http://www.vcu.edu/csbc/bbsi/*
Bioinformatics and computational biology	Cold Spring Harbor Laboratory	*http://www.cshl.edu/URP/nsf~reu*
Computational genomics	Kansas State University	*http://www.kddresearch.org/REU/Summer-2003/announcement.html*
Fungal genomics and computational biology	University of Georgia	*http://www.genetics.uga.edu/undergrad_fgcb.html*
Bioinformatics	Loyola University, Chicago	*http://reu.cs.luc.edu*
Systems biology	Harvard	*http://sysbio.harvard.edu/csb/jobs/undergraduate.html*
Bioinformatics	California State University, Los Angeles	*http://instructional1.calstatela.edu/jmomand2/*
Genomics/bioinformatics	J. Craig Venter Institute	*http://www.jcvi.org/education/internship.php*
Bioinformatics	Greater Philadelphia Bioinformatics Alliance	*http://www.gpba-bio.com/educ_internships.asp*

attributes related to "thinking and working like a scientist," gains in communication and practical skills, and enhanced preparation for graduate school (Seymour et al. 2004). Students also acquire realistic insights into the process of scientific inquiry (Gafney 2001). However, undergraduate research experiences do not appear to attract significant numbers of previously uninterested students to a career that requires a postgraduate degree (Hunter et al. 2006; Lopatto 2004; Seymour et al.

2004). One important caveat pertains to programs that recruit first-year students from underrepresented groups: Such programs might indeed stimulate a student's interest in graduate school, because these students are among those least likely to have had exposure to the idea of graduate school as an option (Seymour et al. 2004). NPGI could promote, and then carefully monitor over time, the expansion of undergraduate research opportunities that result in an expanded and diverse plant genomics community,

Students and faculty members alike cite the importance of dedicated mentoring as a key factor contributing to students' positive responses to their research experience (Lopatto 2003). The primary mentors for many undergraduate research, though, are graduate students and postdoctoral fellows, who may have little or no experience in teaching or mentoring younger scientists, and who could benefit from a recently developed program (Handelsman et al. 2005) that has been validated for effectiveness at 11 institutions (Pfund et al. 2006).

Introductory laboratory courses that engage students in interdisciplinary investigations in plant sciences and genomics are another avenue to promote student interest in research at a time when their career choices are still relatively fluid. Students might assimilate new information more effectively through inquiry-based, collaborative activities than through traditional classroom learning alone (Wood and Gentile 2003). Inquiry-based pedagogical activities in genomics that address significant, novel questions are under development by single institutions (Washington University 2005), by a consortium of small liberal arts colleges working together with Columbia University's Genome Sequencing Center (Carleton College 2007), and by JGI. The Howard Hughes Medical Institute has recently initiated plans for a national genomics research course for undergraduate freshmen. Students from colleges and universities around the country will work collectively on the same research questions, sharing data and results (HHMI 2007). These innovative programs illustrate ways that education can be more fully integrated into NPGI-funded research. **However, it is absolutely vital that already overburdened PIs, or groups of PIs, receive sufficient extra funds, beyond those required to perform their research in an increasingly competitive funding environment, to devote dedicated personnel to these endeavors.** For example, NPGI could consider establishing a new category of PIs dedicated to education, as pioneered by the Howard Hughes Medical Institute through its teaching investigators program. Large plant genomics centers should hire full-time outreach coordinators by appointing professional education managers. Initiatives to be organized at this level could include summer-long funded research internships for community college and high school teachers who wish to develop inquiry-based activities that involve students in the practice of science. Dissemination of information about educational activities deemed successful, as assessed by rigorous outcomes-based metrics, should also be a higher priority for NPGI. One venue for sharing "what works" is to hold sessions devoted to education at the large professional society meetings. By working more closely with the Botanical Society of America and the National Association of Biology Teachers, NPGI can reach thousands of members who teach at the undergraduate and precollege level.

Plant Genomics, National Competitiveness, and International Collaboration

International partnerships like those described in Chapter 2 provide opportunities for U.S. researchers and students to gain valuable experience in a foreign research setting. In an increasingly global scientific arena, U.S. competitiveness will be enhanced by training a cadre of young scientists who understand the advantages of different research environments, the scope of fundamental issues such as food security, and the challenges to national security posed by agricultural constraints. Grand challenge programs that are truly visionary will likely be international in focus, and will require researchers who can design creative and productive programs that are not limited by a single perspective. In this regard, NPGI could seek collaborative funding opportunities with various foundations that are concerned with global agricultural issues, as well other traditional international partners.

The groundwork and personal connections that are needed to help structure successful international research programs are often fostered during the formative years of a scientist's career. Such international research networks harness the creative energy of a young, mobile generation of scientists and the economic power of the emerging economies of Asia (specifically China and India) and of the European Union, the United States, and Australia to provide training, education, and research infrastructure, and to ensure public access to data and information. These considerations suggest a need to increase the opportunity for international training, particularly for our graduate students.

For all of the adopted education recommendations (see below), NPGI should build robust and peer-reviewed methods for assessment. Furthermore, IWG agencies should require all NPGI PIs to report the previous educational background, citizenship, and subsequent career paths for every individual funded by an NPGI grant. NPGI needs to establish a mechanism to collect these data in a centralized location and a set of quantitative criteria by which goals for training can be articulated and measured against this dataset.

Outreach

RECOMMENDATION 9: Promote outreach on plant genomics and related issues that are critical to educating the American public on the value of genomics-based innovations.

Many research programs include components to reach out beyond the scientific community and emphasize the importance of increasing the public understanding of science. Outreach programs in plant genomics are important because end users—food consumers, breeders, farmers, and others—are likely to apply or use products and tools of plant genomic research if they understand the value and benefits of those products and tools, and their potential risks.

As with education, NPGI should build robust and peer-reviewed methods for assessment of any adopted outreach recommendations. Because the goals of such activities will determine the metrics to be used to measure success, the goals of the education and outreach activities have to be clearly defined. For example, if one goal of workshops for K-12 teachers and summer internships for high school students is to broaden the targeted populations' understanding about plant science, genomics, and biotechnology, the conduct of rigorous surveys of participants' knowledge *before* and *after* each program is necessary to assess the impact of the workshop. Longer-term assessment could include occasional follow-up questionnaires to document the broader impact of participation in the workshops on the science curriculum at the teachers' home schools. For programs with an explicit focus on plant biotechnology, student attitudes about biotechnology could also be monitored before and after the activity or internship.

Likewise, one common approach to K-12 outreach in science education is a short-term classroom visit or series of visits by a researcher. The visiting researchers might lead a hands-on activity or talk with the students about societal implications of their research. The goals of such visits are to generate enthusiasm among students for science, improve the image of scientists, and promote science literacy (Laursen et al. 2007). There is little direct evidence that classroom visits achieve those goals. One qualitative assessment of a best-case "scientist in the classroom" program documented some measures of success and several benefits and some potential costs to the graduate students who participate in the program (Laursen et al. 2007).

Scientists who wish to develop an outreach program might not know how to do so effectively and might be unfamiliar with existing resources that could guide them and prevent unnecessary duplication of efforts. In the face of increasing time pressures on principal investigators, graduate students, and postdoctoral fellows, there is little sense in researchers "reinventing the wheel" with respect to outreach and pre-college education (Dolan et al. 2004). **Additional support from NPGI for personnel explicitly trained in outreach who help PIs and graduate students to define, achieve, and further their outreach goals, including outreach to extension and breeder groups, is critical for the translation of NPGI science into tangible benefits to society.**

As has been observed (Labov 2006), "the kinds of experiences (or lack thereof)

BOX 3-2
The Partnership for Research and Education in Plants

The Partnership for Research and Education in Plants, Biology (PREP; Virginia Polytechnic Institute and State University 2006), funded by NSF's Arabidopsis 2010 Project, NIH, and the American Society for Plant Biologists, allows high school students to contribute to real research projects. PREP has involved over 10,000 students, 54 teachers, and 26 scientists in six states. It is the brainchild of a biology teacher, a plant geneticist, and a faculty-level outreach coordinator working together (Dolan et al. 2004). High school students design experiments to characterize novel *Arabidopsis* mutants. The students collect and analyze data on growth and development of the plant lines, and report the data in an online notebook that facilitates interactions with the partner researchers (peers and professional scientists). PREP exemplifies at least three of the four principles of instructional design advocated by a recent NRC report on successful laboratory exercises: 1) they are designed with clear learning outcomes in mind; 2) they integrate the learning of science content with learning about the process of science; and 3) they incorporate ongoing student reflection and discussion (NRC 2005b).

in science that students encounter during their K-12 years will have direct consequences on what college-level instructors will be able to accomplish in their own classrooms and teaching laboratories." By joining forces to expand implementation of an existing program such as the Partnership for Research and Education in Plants (Box 3-2), rather than cobbling together a forced activity lacking a well-considered rationale, NPGI investigators could have a national impact on high school education.

Bringing Genomics to the Sustainable, Local, and Organic Agriculture Communities

The communities of small-scale and organic farmers are expanding in both numbers and in economic and political clout, driven by rising consumer demand for sustainable and locally grown food. This market sector is likely to grow, especially if food transportation costs rise dramatically. Philosophical interest in plant genomics among these groups is likely to benefit from clear communication that genomic research is not necessarily tied to deployment of transgenic plants, and on tangible and relevant outcomes in the form of cultivars that are well suited for particular, local, and often low technological input, agricultural niches.

The committee suggests that NPGI investigate creative mechanisms to translate its research into benefits for such growers. An example is the Public Seed Initiative at Cornell (Cornell University 2005), whose focus is on developing, maintaining, and distributing seeds for cultivars of fruits, vegetables, and grains adapted to the needs of organic and local fresh-market growers. By continuing its current focus on applications of genomic tools to marker-assisted selection for diverse crops and traits, NPGI will also provide useful information to this grower community. For example, both large- and small-scale farmers benefit from markers that speed the development of cultivars with enhanced disease resistance. In addition, **NPGI is poised to apply technological advances in metagenomics, metabolomics, and systems biology** to characterize the complex interdependences among species that are considered important to various cultivation systems, including organic. For example, identification of microbial population structures and the nature of metabolites found in disease suppressive soil ecosystems will help to guide the development of agronomic practices that reduce the need for pesticide use. One interesting case could be investigation of permacultures. Permacultures have minimal need for fertilization, irrigation, or pesticide usage because ecological processes common to forestry ecosystems or agroecosystems are used to maximize the yield of edible species in perennial agroforestry and polyculture systems (Jacke 2005; Mollison 1988).

To identify specific traits that are needed in new food crop cultivars, NPGI could take steps to engage small-scale farmers and relevant trade groups and to facilitate direct interactions between farmers, breeders, and researchers. This could entail interactions between genome scientists and producers to address genomics applications germane to this arena, and participation from producers to identify what portions of the genomics toolkit are most relevant to them and with respect to what traits, at extension events, county fairs, and local farmers' markets.

Since the officially sanctioned organic farming community has banned applications involving human directed recombinant DNA manipulations (for example, genetically modified organisms [GMO]; AMS 2007), even the most well-meaning efforts to create common ground between genomic researchers and organic farmers could be derailed by negative grower or public perceptions that simplistically equate plant genomics with genetically engineered plants and/or proprietary technologies owned by multinational corporations. By fostering a climate of enhanced understanding and promoting connections among researchers, farmers, small-scale seed producers, and nonprofit organizations, NPGI researchers might pave the way for acceptance among small scale growers of a variety of plant genomics technologies.

Examples of a technology that NPGI might seek to further develop and communicate as part of this effort are improved forms of "all-native" or "cisgenic" transformation (discussed earlier in this chapter). These basic approaches need to be made efficient in a variety of species, improved so that mutagenesis during gene transfer is minimized, and made more precise via gene targeting and allele replacement capability. They also need to be publicly accessible (that is, not dominated by private sector patents) so that localized, plant variety- and region-specific use is feasible.

Ethical, Legal, and Social Issues in Plant Genomics

In the Human Genome Project (HGP), ethical, legal, and social issues (ELSI) related to human genomic data and related technologies were recognized to have major impacts on how genomic information would be used in biomedicine. ELSI research at the National Human Genome Research Institute (NHGRI) began in 1990 to understand the social implications of genetic and genomic research. Its orientation has been consciously proactive, in that it seeks to identify "... problem areas ... and solutions ... before scientific information is integrated into health care practice" (NHGRI 2007). The ELSI program accounts for more than $18 million of the annual $485 million HGP budget.

In contrast, there has been little ELSI-related activity in plant genomics research. Despite initial plans to the contrary (NSTC 1998) and reemphasis (NSTC 2000), only a narrower objective was retained under broader impacts of the NPGI, stating that "research is needed to identify methods for more effective communication with the general public" (NSTC 2003). To date, there has not been significant collaborative engagement with social scientists to conduct scholarly research on the causes and resolution of ELSI issues related to plant genomics (NSTC 1999, 2000, 2001, 2003, 2004, 2005, 2006, 2007). The lack of attention to ELSI programs from the NPGI is surprising in that it comes amidst growing controversies in the agricultural, forestry, and energy sectors about genetic technologies, particularly transgenic approaches, and in a political climate where public skepticism regarding the economics associated with government-subsidized ethanol production are becoming ever more important.

Because of attendant costs and social controversies, issues that could be addressed via ELSI research have effectively removed GMO tools for translation of genomic knowledge into useful products from all but the largest commodity crops and the largest agricultural companies, and in only a subset of countries. GMOs have served as a focal point for analysis of a large number of ELSI issues that are growing in significance for agriculture (Serageldin 1999); these may logically spread to encompass all of genomics-enabled breeding in the future.

The limited attention to ELSI issues by NPGI may have impacted public perception of plant genomics and associated biotechnologies. In the acrimonious GMO debate, most of the NPGI-funded genomics research community has been conspicuously quiet, even when the debate concerns substantive genomics issues. This may have helped to create space for those with strong political views, but weak knowledge of plant science, to dominate the social discourse (Vasil 2003), promoting confusion on the part of the media and public. The plant genomics community could provide context for understanding the impacts of GMOs, in comparison to the effects of accepted practices of breeding and domestication, on plant genomes. **Because fundamental advances in knowledge of plant genomes are likely to empower increasingly novel, innovative uses of genomic information, the opportunity cost to society from its limited ability to use transgenic approaches is likely to grow rapidly.**

Outreach on ELSI topics is an issue that the NPGI needs to confront. A next generation of teachers and scientists who are trained in both plant genomics and ELSI issues could contribute to resolution of genomics-related social issues, and thus play a valuable role in guiding the development of scientifically sound regulations. The outcome could have profound consequences for deployment of the products of plant genomics, and on laws that govern international trade. Potential ELSI issues of interest to NPGI include those listed in Box 3-3.

Goals for Education and Outreach (Recommendations 8 and 9)

5-year goals

- Develop evidence-based metrics to assess educational and outreach programs.
- Enhance opportunities for graduate and undergraduate students to become proficient in the theory as well as the practice of computational biology and bioinformatics, through graduate fellowships and undergraduate research experiences.
- Establish interdisciplinary graduate postdoctoral fellowships in plant genomics with an option for international collaborations. This could be modeled on the success of the *Arabidopsis* 2010 Project: International Research Experience for Graduate Students and Postdoctoral Fellows that supports exchanges between U.S. and German laboratories.
- Stimulate undergraduate student interest in plant genomics, especially among populations of students who might be less aware of research career opportunities, through expanded research opportunities with trained mentors and through integrated inquiry-based activities in undergraduate and precollege courses.
- Develop well-designed educational activities that draw on the latest learning theory research and devise mechanisms for educators to share these initiatives, by creating a new class of PIs dedicated to education and by funding professional education managers to coordinate outreach activities.
- Expand the Plant Genome Research Outreach Portal (PGROP) to include a comprehensive collection of existing outreach programs, with evaluative information, and links to assessment tools.

BOX 3-3
Ethical, Legal, and Social Issues Associated with Plant Genomics

- Biological and social benefits and risks of plant modifications derived from genomics that use recombinant DNA, such as genetic engineering or GMO approaches.
- Sustainability and biodiversity in the broad biological and social senses, and the extent to which genomics-aided breeding can aggravate or mitigate these concerns.
- The appropriate role for formal social controls on intellectual property protection and regulation as related to income distribution, business development, and social cost and benefit tradeoffs.
- The increasing controls over international germplasm movement and attendant concerns about biopiracy. These regulations may seriously hinder genomics based plant breeding and research progress in both the developed and developing worlds.
- The extent and control of unintentional contamination of germplasm. Critical translational research can be hampered—against a backdrop of stringent social intolerance—by dispersal and adventitious presence of foreign genes due to inadvertent pollen, seed, and vegetative dispersal from exotic genotypes, species, and transgenes.
- Public education and outreach about the goals and rationale for genomics and related biotechnology research. Broad social approval of plant genomics deployment will be dependent on judgments of social and personal benefit in comparison to risk. (Hossain et al. 2003).

- NPGI PIs should try to forge connections with engineers and computational scientists, with the goal of attracting students in these fields to plant genomics at the graduate level.
- NPGI PIs should encourage changes in the undergraduate curriculum at their own institutions and participate in the reformation. PIs should also be encouraged to participate in similar reforms in their institutional Ph.D. programs in genomics so that two courses in statistics and competence in a modern scripting language become standard requirements for advanced degrees.
- Establish mechanisms to engage sustainable, organic, and small-scale farmers in identification of specific traits for which applications of genomic tools could lead to usable varieties with enhanced performance characteristics.

10-year goal

- Expand training in ethical, legal, and social issues pertaining to plant genomics for K-12 and undergraduate students and teachers, and for NPGI predoctoral and postdoctoral stipend recipients.

20-year achievements

- Integration of research and education in plant genomics will rival that of biomedical genomics in creativity, in public profile, and in the ability to attract new students.

The plant genomics community will provide leadership in contributions toward public outreach on ELSI issues, including engagement in development of science-based regulatory policies at national and international levels, by NPGI-funded programs and NPGI-trained students and postdoctoral associates.

Glossary

Amplified fragment length polymorphism	A highly sensitive polymerase chain reaction for detecting polymorphisms in DNA.
Annotation	The process of adding an explanatory note or a body of notes to a biological subject. For example, sequencing annotation involves identifying the locations and coding regions of genes on a sequence. In functional genomics, annotation is the determination of what various gene products do. In mutational studies, annotation is the recording of observed phenotypes.
Artificial micro-RNA	A transgene engineered to produce a micro RNA targeted against a particular DNA sequence.
Bacterial artificial chromosome	Vectors used to clone large (100-300 kb) DNA fragments.
Bioinformatics	The study of genetic and other biological information using computer, mathematical, and statistical techniques.

cDNA	Complementary DNA; made from messenger RNA (mRNA).
Commensals	Two organisms of different species in a relationship that benefits one but does not affect the other species.
Common query language	A formal language for representing queries to information retrieval systems.
Comparative genomics	The comparison of gene and genome structure, function, and evolution across taxa.
DNA fingerprinting	An analysis used to identify an individual's unique sequence of DNA base pairs by exposing the individual's DNA sample to molecular probes.
DNA marker	A DNA sequence in a known location of a chromosome that has a particular trait.
Epigenetics	The study of heritable changes in gene function that occur without a change in the DNA sequence.
Epigenomics	The study of alterations to the genome DNA sequence that alters gene function without a change in DNA sequence.
Expressed gene content	The sum of all mRNAs expressed in a given sample.
Expressed sequence tag (ESTs)	The result of large-scale partial sequencing of randomly selected cDNA clones. ESTs are a useful tool for gene identification, localization, and mapping.
Extensible markup language	A general purpose language used to describe data in a platform-neutral way so that data can be shared across multiple platforms.
Functional genomics	The study of genomes to determine the biological function of all the genes and their products.
Gene expression	The transcription, translation, and phenotypic manifestation of a gene.

Gene Ontology	The Gene Ontology project provides a controlled vocabulary to describe gene and gene product attributes in any organism. Read more about the Gene Ontology at geneontology.org.
Genetic polymorphism	Different forms of a genetic sequence that arise from mutation.
Genetic redundancy	The overlapping of function of two or more genes that results in no change in the biological phenotype if one of those genes is inactivated.
Genome	The entire chromosomal genetic material of an organism.
Genome sequencing	Determination of the order of nucleotides (base sequences) of an entire genome.
Genomics	The comprehensive study of whole sets of genes and their interactions rather than single genes or proteins.
Germplasm	The genetic material that carries the inherited characteristics of an organism.
High throughput	Rapid (and simultaneous) processing of large sample sets.
Informatics	The study of the application of computer and statistical techniques to the management of information (also see bioinformatics).
Kinase	An enzyme that adds phosphate groups to proteins.
Marker-assisted breeding	The use of DNA polymorphisms as a means to rapidly test progeny from a cross for the introgression of a desired chromosomal segment.
Metabolomics	The global analysis of metabolites and metabolic networks in cells, tissues, and organ systems.
Metadata	Data that describe other data.

Metagenome	The collective genome of a community of microorganisms, obtained without first isolating individual community members.
Metagenomics	The study of genetic material recovered directly from environmental samples.
Microarray	A microscope slide or other solid support on which many distinct cDNAs or DNA oligonucleotides are patterned at high density in an addressable array. Microarrays are interrogated by hybridization to fluorescently labeled cDNAs or RNAs to detect the genes that are actively transcribed.
Microsatellite	A polymorphic sequence of DNA consisting of tandemly repeated units of DNA where the repeat unit is usually 1-4 nucleotides long.
mRNA	Messenger RNA.
Mutagenesis	The development of a mutation.
Mutualists	Organisms that are associated with each other in a mutually beneficial relationship.
Open reading frame	A reading frame in a sequence of nucleotides in DNA that contains no termination codons and so can potentially translate as a polypeptide chain.
Organogenesis	The development of organs in living organisms.
Phenomics	The study of the full complement of phenotypic traits expressed as a result of genetic and environmental influences.
Phylogeny	Evolutionary relationships among organisms; the developmental history among organisms.
Ploidy	A multiple of the basic number of chromosomes in a cell.

Positional cloning	The isolation of a gene based solely on its position along a chromosome and its mutant of epigenetic phenotypic alteration.
Protease	An enzyme that hydrolyzes proteins.
Protein interactome	The full complement of protein-protein interaction in a cell.
Proteomics	The study of a full set of proteins encoded by a genome.
Putative unique transcript	An assembled complementary DNA (cDNA) sequence generated by sampling many overlapping expressed sequence tags.
Quantitative trait loci	Genetic loci that affect a quantitatively inherited trait.
Reference sequence	An accurately assembled genomic sequence that can serve as a reference for comparative genomic studies.
Rhizosphere	The zone around plant roots.
Shared descriptive language	A computer language used in annotation to express a quality of the biological subject of interest, for example the "transcribed region" of this sequence. The annotators in the genomics community jointly use the terms or words in this common language to facilitate comparisons of data.
Single-nucleotide polymorphism	DNA sequence variation that occurs when a single nucleotide (A, T, G, or C) is altered. SNPs can be useful in detecting genetic variation among individuals in a given population.
Splicing	The process that removes intron sequences and covalently joins exon sequences.
Structured query language	A standard interactive and programming language for getting information from and updating a database.
Symbionts	Organisms that are associated with each other in a mutually beneficial relationship.

T-DNA	Transferred DNA from Agrobacterium.
Targeting Induced Local Lesions in Genomes (TILLING)	A method for identifying mutations in a specific target gene.
Transcriptomics	The study of the full complement of activated genes, transcripts, and mRNA in a given tissue at any particular time.
Translational genomics	Translation of genomic findings from experimental model species to crop species
Transposon	A segment of DNA that moves, jumps, and can be integrated into many different sites along a chromosome.

References

AFPA (American Forest and Paper Association). 2007. U.S. Forest Products Industry: Competitive Challenges in a Global Marketplace. Accessed on September 19, 2007 at *http://policycouncil.nationaljournal.com/EN/Forums/American+Forest++Paper+Association/1f7d98e5-d352-45be-8ec6-0d335ee9a4f7.htm.*

AMS (U.S. Department of Agriculture, Agricultural Marketing Service). 2007. The National Organic Program. Accessed on October 5, 2007 at *http://www.ams.usda.gov/NOP/NOP/NOPhome.html.*

ARS (U.S. Department of Agriculture, Agricultural Research Service). 2005. National Plant Germplasm System. Accessed on August 13, 2007 at *http://www.ars-grin.gov/npgs/index.html.*

Baran S.B., C.J. Lawrence, and V. Brendel. 2004. Plant genome research outreach portal: a gateway to plant genome research "outreach" programs and activities. *Plant Physiology* 134:889.

Bialek, W., and D. Botstein. 2004. Introductory science and mathematics education for 21st-century biologists. *Science* 303:788–790.

Boyer, J.S. 1982. Plant productivity and environment. *Science* 218:443–448.

Broad Institute. 2007. Fungal Genome Initiative. Accessed on November 16, 2007 at *http://www.broad.mit.edu/annotation/fgi/.*

Brown, G.R., G.P. Gill, R.J. Kuntz, C.H. Langley, and D.B. Neale. 2004. Nucleotide diversity and linkage disequilibrium in loblolly pine. *Proceedings of the National Academy of Sciences USA* 101:15255–15260.

Brueggeman, R., N. Rostoks, D. Kudrna, A. Kilian, F. Han, J. Chen, A. Druka, B. Steffenson, and A. Kleinhofs. 2002. The barley stem rust-resistance gene Rpg1 is a novel disease-resistance gene with homology to receptor kinases. *Proceedings of the National Academy of Sciences USA* 99:9328–9333.

Brunner, A.M., V.B. Busov, and S.H. Strauss. 2004. Poplar genome sequence: functional genomics in an ecologically dominant plant species. *Trends in Plant Science* 9:49–56.

Calhill, D.J., and D.H Schmidt. 2004. Use of DNA marker assisted selection in a product development breeding program. In *Proceedings of the 4th International Crop Science Conference*, 26 Sept.–1 Oct. 2004. Brisbane, Australia. Available at *http://www.cropscience.org.au/icsc2004/symposia/3/4/133_schmidtdh.htm#TopOfPage.*

Campbell, A.M., and L.J. Heyer. 2007. *Discovering Genomics, Proteomics, and Bioinformatics.* 2nd ed. San Francisco: Benjamin Cummings.

Campbell, A.M., M.L. Ledbetter, L.L. Hoopes, T.T. Eckdahl, L.J. Heyer, A. Rosenwald, E. Fowlks, S. Tonidandel, B. Bucholtz, and G. Gottfried. 2007. Genome consortium for active teaching: meeting the goals of BIO2010. *CBE—Life Sciences Education* 6:109–118.

Carleton College. 2007. Teaching Big Science at Small Colleges: A Genomics Collaboration. Accessed on August 23, 2007 at *http://serc.carleton.edu/genomics/index.html.*

CGIAR (Consultative Group on International Agricultural Research) Genomics Task Force. 2006. Enhancing the delivery of genomics research outcomes. Accessed on October 30, 2007 at *http://www.sciencecouncil.cgiar.org/meetings/meeting/SC5/SC5%20Item%206%20Genomics.pdf.*

Chapple, C., M. Ladisch, and R. Meilan. 2007. Loosening lignin's grip on biofuel production. *Nature Biotechnology* 25:746–748.

Check, E. 2003. Harvard heralds fresh take on systems biology. *Nature Biotechnology* 425:439.

Clark, R.M., G. Schweikert, C. Toomajian, S. Ossowski, G. Zeller, P. Shinn, N. Warthmann, T.T. Hu, G. Fu, D.A. Hinds, H. Chen, K.A. Frazer, D.H. Huson, B. Scholkopf, M. Nordborg, G. Ratsch, J.R. Ecker, and D. Weigel. 2007. Common sequence polymorphisms shaping genetic diversity in *Arabidopsis thaliana. Science* 317:338–342.

Cornell University. 2005. Public Seed Initiative. Accessed on October 5, 2007 at *http://www.plbr.cornell.edu/psi/index.html.*

Davidson College. 2005. Genome Consortium for Active Teaching (GCAT). Accessed on August 10, 2005 at *http://www.bio.davidson.edu/projects/gcat/gcat.html.*

Delmer, D.P. 2005. Agriculture in the developing world: connecting innovations in plant research to downstream applications. *Proceedings of the National Academy of Sciences USA* 102:15739–15746.

Distelfeld, A., C. Uauy, T. Fahima, and J. Dubcovsky. 2006. Physical map of the wheat high-grain protein content gene Gpc-B1 and development of a high-throughput molecular marker. *New Phytologist* 169:753–763.

DOE (U.S. Department of Energy). 2005a. Energy and environmental profile of the pulp and paper industry. Accessed on September 19, 2007 at *http://www1.eere.energy.gov/industry/forest/pdfs/pulppaper_profile.pdf.*

———. 2005b. Systems biology for energy and environment. In *DOE Genomics: GTL Roadmap*: U.S. Department of Energy Office of Science. Accessed on October 30, 2007 at *http://genomicsgtl.energy.gov/roadmap/pdf/GenomicsGTL_Roadmap_lowres.pdf.*

———. 2006. Breaking the Biological Barrier to Cellulosic Ethanol: A Joint Research Agenda, DOE/SC-0095, U.S. Department of Energy Office of Science of Energy Efficiency and Renewable Energy. Accessed on September 19, 2007 at *http://genomicsgtl.energy.gov/biofuels/2005workshop/b2blowres63006.pdf.*

———. 2007a. Plant Feedstocks Genomics for Bioenergy. Accessed on October 22, 2007 at *http://genomicsgtl.energy.gov/research/DOEUSDA/index.shtml.*

———. 2007b. Community Sequencing Plans. Accessed on July 31, 2007 at *http://www.jgi.doe.gov/sequencing/seqplans.html.*

Dolan, E.L., B.E. Soots, P.G. Lemaux, S.Y. Rhee, and L. Reiser. 2004. Strategies for avoiding reinventing the precollege education and outreach wheel. *Genetics* 166:1601–1609.

Doty, S.L., T.Q. Shang, A.M. Wilson, J. Tangen, A.D. Westergreen, L.A. Newman, S.E. Strand, and M.P. Gordon. 2000. Enhanced metabolism of halogenated hydrocarbons in transgenic plants containing mammalian cytochrome P450 2E1. *Proceedings of the National Academy of Sciences USA* 97:6287–6291.

Dyer, B.D., and M.D. LeBlanc. 2002. Meeting report: incorporating genomics research into undergraduate curricula. *Cell Biology Education* 1:101–104.

Easterling, W.E., P.K. Aggarwal, P. Batima, K.M. Brander, L. Erda, S.M. Howden, A. Kirilenko, J. Morton, J-F. Soussan, J. Schmidhuber, and F.N. Tubiello. 2007. Food, fiber, and forest products. In *Climate Change 2007: Impacts, Adaptation and Vulnerability. Contribution of Working Group II to the Fourth Assessment of the Intergovernmental Panel on Climate Change*, edited by M.L. Parry, O.F. Canziani, J.P. Palutikof, P.J.v.d. Linden, and C.E. Hanson. Cambridge: Cambridge University Press.

Endy, D. 2005. Foundations for engineering biology. *Nature* 438:449–453.

ERS (U.S. Department of Agriculture, Economic Research Service). 2006. Forest Land. Accessed on October 30, 2007 at *http://www.ers.usda.gov/Data/MajorLandUses/glossary.htm#forest.*

ERS (U.S. Department of Agriculture, Economic Research Service). 2007. U.S. Agricultural Projections to 2016. Accessed on November 16, 2007 at *http://www.ers.usda.gov/publications/oce071/oce20071.pdf.*

European Commission. 2006. EU-SOL - High Quality Solanaceous Crops for Consumers, Processors and Producers by Exploration of Natural Biodiversity. Accessed on August 24, 2007 at *http://www.eu-sol.net.*

FAOSTAT. 2007. Core database of the Food and Agriculture Organization of the United Nations. Accessed on May 23, 2007 at *http://faostat.fao.org/default.aspx*.

Fernandez-Cornejo, J, and M. Caswell. 2006. *The First Decade of Genetically-Engineered Crops in the United States*. U.S. Department of Agriculture Economic Research Service, April 2006. Available from *http://www.ers.usda.gov/publications/eib11/eib11.pdf*.

Fu, D., P. Szucs, L. Yan, M. Helguera, J.S. Skinner, J. von Zitzewitz, P.M. Hayes, and J. Dubcovsky. 2005. Large deletions within the first intron in VRN-1 are associated with spring growth habit in barley and wheat. *Molecular Genetics and Genomics* 273:54–65.

Gafney, L. 2001. The impact of research on undergraduates' understanding of science. *Council on Undergraduate Research Quarterly* 21:172–176.

Gerstein, M., D. Greenbaum, K. Cheung, and P. L. Miller. 2007. An interdepartmental Ph.D. program in computational biology and bioinformatics: the Yale perspective. *Journal of Biomedical Informatics* 40(1):73–79.

GMOD (Generic Model Organism Database). 2007. GMOD Home Page. Accessed on September 20, 2007 at *http://gmod.org/*.

Goff, S.A., and H.J. Klee. 2006. Plant volatile compounds: sensory cues for health and nutritional value? *Science* 311:815–819.

Gramene. 2007. RiceCyc Home and Biochemical Pathways. Accessed on September 28, 2007 at *http://www.gramene.org/pathway/*.

Handelsman, J., C. Pfund, S. Miller Lauffer, and C. Pribbenow. 2005. *Entering Mentoring: A Seminar to Train a New Generation of Scientists*. Madison: University of Wisconsin Press.

HHMI (Howard Hughes Medical Institute). 2007. *HHMI pilots a "SEA" change: Experimental course to bring genomics teaching and research to undergraduates across the nation*. Accessed on August 23, 2007 at *http://www.hhmi.org/bulletin/aug2007/chronicle/sea.html*.

Honts, J.E. 2003. Evolving strategies for the incorporation of bioinformatics within the undergraduate cell biology curriculum. *Cell Biology Education* 2(4):233–247.

Hossain, F., B. Onyango, B. Schiling, and W. Hallman. 2003. Public perceptions of biotechnology and acceptance of genetically modified food. *Journal of Food Distribution Research* 4:36–50.

Huang, J., R. Hu, S. Rozelle, and C. Pray. 2005. Insect-resistant GM rice in farmers' fields: assessing productivity and health effects in China. *Science* 308 (5722):688–690.

Huang, J., S. Rozelle, C. Pray, and Q. Wang. 2002. Plant biotechnology in China. *Science* 295(5555):674–676.

Hunter, A.-B., S.L. Laursen, and E. Seymour. 2006. Becoming a scientist: the role of undergraduate research in students' cognitive, personal, and professional development. *Science Education* 91:36–74.

Ideker, T. 2004. Systems biology 101–what you need to know. *Nature Biotechnology* 22:473–475.

IRFGC (International Rice Functional Genomics Consortium). 2007. Web 2.0 Portal. Accessed on August 9, 2007 at *http://www.iris.irri.org.8080/IRFGC/*.

IRGSP (International Rice Genome Sequencing Project). 2002. Completion of the rice genome sequence announced. Accessed on October 30, 2007 at *http://rgp.dna.affrc.go.jp/E/IRGSP/Dec18_NEWS.html*.

Jacke, D. 2005. *Edible Forest Gardens: Ecological Vision and Theory for Temperate Climate Permaculture*. White River Junction, VT: Chelsea Green Publishing.

JGI (U.S. Department of Energy, Joint Genome Institute). 2006. Energy-rich Portfolio of New Genome Sequencing Targets for DOE JGI. Accessed on August 10, 2007 at *http://www.jgi.doe.gov/News/news_7_11_06.html*.

Johnson, G.R., N.C. Wheeler, and S.S. Strauss. 2000. Financial feasibility of marker-aided selection in Douglas-fir. *Canadian Journal of Forest Research* 30:1942–1952.

Kim, S., V. Plagnol, T.T. Hu, C. Toomajian, R.M. Clark, S. Ossowski, J.R. Ecker, D. Weigel, and M. Nordborg. 2007. Recombination and linkage disequilibrium in Arabidopsis thaliana. *Nature Genetics* 39(9):1151–1155.

Labov, J. 2006. Education at the National Academies: three recent reports on improving science education. *Cell Biology Education* 5:12–18.

Laursen, S., C. Liston, H. Thiry, and J. Graf. 2007. What good is a scientist in the classroom? Participant outcomes and a program design features for a short-duration science outreach intervention in K-12 classrooms. *Cell Biology Education* 6:49–64.

Lopatto, D. 2003. The essential features of undergraduate research. *Council on Undergraduate Education Quarterly* 23:139–142.

———. 2004. Survey of undergraduate research experiences (SURE): first findings. *Cell Biology Education* 3:270–277.

Meyerowitz, E., and S.Y. Rhee. 2006. Report of the NSF Workshop for a Plant Cyberinfrastructure Center. October 17–18, 2005. Accessed on August 10, 2007 at *http://www.nsf.gov/pubs/reports/plant_cyberin_report_20060322.pdf.*

Mollison, B. 1988. *Permaculture: A Designer's Manual.* Tyalgum, Australia: Tagari Publications.

NASS (U.S. Department of Agriculture, National Agricultural Statistics Service). 2007a. Prospective Plantings. Accessed on September 19, 2007 at *http://usda.mannlib.cornell.edu/usda/current/ProsPlan/ProsPlan-03-30-2007.pdf.*

———. 2007b. Crop values 2006 summary. Accessed on September 19, 2007 at *http://usda.mannlib.cornell.edu/usda/current/CropValuSu/CropValuSu-02-15-2007.pdf.*

———. 2007c. USDA National Agricultural Statistics Service - Quick Stats. Accessed on November 16, 2007 at *http://www.nass.usda.gov/QuickStats/Create_Federal_All.jsp.*

Neale, D.B., and O. Savolainen. 2004. Association genetics of complex traits in conifers. *Trends in Plant Science* 9:325–330.

NHGRI (National Human Genome Research Institute). 2007. *ELSI Planning and Evaluation History.* Accessed on October 5, 2007 at *http://www.genome.gov/10001754.*

NRC (National Research Council). 2002. *The National Plant Genome Initiative: Objectives for 2003–2008.* Washington, DC: The National Academies Press.

———. 2003. *BIO 2010. Transforming Undergraduate Education for Future Research Biologists.* Washington, DC: The National Academies Press.

———. 2005a. *Review of the Department of Energy's Genomics: GTL Program.* Washington, DC: The National Academies Press.

———. 2005b. *America's Lab Report: Investigations in High School Science.* Washington, DC: The National Academies Press.

———. 2007a. *The New Science of Metagenomics. Revealing the Secrets of Our Microbial Planet.* Washington, DC: The National Academies Press.

———. 2007b. *Rising Above the Gathering Storm.* Washington, DC: The National Academies Press.

NSF (National Science Foundation). 2006. REU Sites: Biological Sciences. Accessed on August 10, 2007 at *http://www.nsf.gov/crssprgm/reu/list_result.cfm?unitid=5047.*

———. 2007a. 2010 Project. Accessed on November 14, 2007 at *http://www.nsf.gov/funding/pgm_summ.jsp?pims_id=5337&org=BIO&sel_org=BIO&from=fund.*

———. 2007b. Developing Country Collaborations in Plant Genome Research (DCC-PGR). Accessed on August 1, 2007 at *http://www.nsf.gov/pubs/2004/nsf04563.htm.*

———. 2007c. NSF Awards Made to Support the Construction of BAC Libraries. Accessed on October 5, 2007 at *http://www.nsf.gov/bio/pubs/awards/bachome.htm.*

NSTC (National Science and Technology Council). 1998. National Plant Genome Initiative. Washington, DC: National Science and Technology Council.

———. 1999. National Plant Genome Initiative: Progress Report October 1999. Washington, DC: National Science and Technology Council.

———. 2000. National Plant Genome Initiative: Progress Report November 2000. Washington, DC: National Science and Technology Council.

———. 2001. National Plant Genome Initiative: Progress Report December 2001. Washington, DC: National Science and Technology Council

———. 2003. National Plant Genome Initiative: 2003–2008. Washington, DC: National Science and Technology Council

———. 2004. National Plant Genome Initiative: Progress Report January 2004. Washington, DC: National Science and Technology Council

———. 2005. National Plant Genome Initiative: Progress Report January 2005. Washington, DC: National Science and Technology Council

———. 2006. National Plant Genome Initiative: Progress Report January 2006. Washington, DC: National Science and Technology Council

———. 2007. National Plant Genome Initiative: Progress Report January 2007. Washington, DC: National Science and Technology Council.

Pevzner, P.A. 2004. Educating biologists in the 21st century: bioinformatics scientists versus bioinformatics technicians. *Bioinformatics* 20:2159–2161.

Pfund, C., C. Maidl Pribbenow, J. Branchaw, S. Miller Lauffer, and J. Handelsman. 2006. Professional skills: the merits of training mentors. *Science* 311:473–474.

PGROP (Plant Genome Research Outreach Portal). 2006. Plant Genome Research Outreach Portal. Accessed on August 12, 2007 at *http://plantgdb.org/PGROP/pgrop.php.*

Plant Genome Database. 2007. Resources for Comparative Genomics. Accessed on September 20, 2007 at *http://www.plantgdb.org/prj/ESTCluster/progress.php.*

Qaim, M., and D. Zilberman. 2003. Yield effects of genetically modified crops in developing countries. *Science* 299:900–902.

Raboy, V. 2007. The ABCs of low-phytate crops. *Nature Biotechnology* 25:874–875.

Reddy, K.R., and H.F. Hodges. (eds.) 1999. *Climate Change and Global Crop Productivity.* Wallingford and New York: CAB International.

Rommens, C.M. 2007. Intragenic crop improvement: combining the benefits of traditional breeding and genetic engineering. *Journal of Agricultural and Food Chemistry* 55:4281-4288.

Scientific Inference Systems Laboratory. 2007. The Computable Plant. Accessed on September 28, 2007 at *http://www.computableplant.org/.*

Serageldin, I. 1999. Biotechnology and food security in the 21st century. *Science* 285:387–389.

Seymour, E., A.B. Hunter, S.L. Laursen, and T. Deantoni. 2004. Establishing the benefits of research experiences for undergraduates in the sciences: first findings from a three-year study. *Science Education* 88:493–534.

Somerville, C. 2006. The billion-ton biofuel vision. *Science* 312:1277.

Strauss, S.H., R. Lande, and G. Namkoong. 1992. Limitations of molecular-marker-aided selection in forest tree breeding. *Canadian Journal of Forest Research* 22:1050–1061.

TAIR (The Arabidopsis Information Resource). 2007. AraCyc 4.0. Accessed on September 28, 2007 at *http://www.arabidopsis.org/biocyc/.*

The Arabidopsis Genome Initiative. 2000. Analysis of the genome sequence of the flowering plant *Arabidopsis thaliana. Nature* 408:796–815.

Traxler, G. 2004. The Economic Impacts of Biotechnology-based Technological Innovations: ESA Working Paper No. 04-08. Rome: Food and Agriculture Organization of the United Nations.

Trenberth, K.E., P.D. Jones, P. Ambenje, R. Bojariu, D. Easterling, A. Klein Tank, D. Parker, F. Rahimzadeh, J.A. Renwick, M. Rusticucci, B. Soden, and P. Zhai. 2007. Observations: surface and atmospheric climate change. In *The Physical Science Basis. Contribution of Working Group 1 to the Fourth Assessment Report of the Intergovernmental Panel on Climate Change,* edited by S. Solomon, D. Qin, M. Manning, Z. Chen, M. Marquis, K.B. Averyt, M. Tignor, and H.L. Miller. New York: Cambridge University Press.

Tuberosa, R., and S. Salvi. 2006. Genomics-based approaches to improve drought tolerance of crops. *Trends in Plant Science* 11:405–412.

Tuskan, G.A., S. DiFazio, S. Jansson, J. Bohlmann, I. Grigoriev, U. Hellsten, N. Putnam, S. Ralph, S. Rombauts, A. Salamov, J. Schein, L. Sterck, A. Aerts, R.R. Bhalerao, R.P. Bhalerao, D. Blaudez, W. Boerjan, A. Brun, A. Brunner, V. Busov, M. Campbell, J. Carlson, M. Chalot, J. Chapman, G.L. Chen, D. Cooper, P.M. Coutinho, J. Couturier, S. Covert, Q. Cronk, R. Cunningham, J. Davis, S. Degroeve, A. Dejardin, C. dePamphilis, J. Detter, B. Dirks, I. Dubchak, S. Duplessis, J. Ehlting, B. Ellis, K. Gendler, D. Goodstein, M. Gribskov, J. Grimwood, A. Groover, L. Gunter, B. Hamberger, B. Heinze, Y. Helariutta, B. Henrissat, D. Holligan, R. Holt, W. Huang, N. Islam-Faridi, S. Jones, M. Jones-Rhoades, R. Jorgensen, C. Joshi, J. Kangasjarvi, J. Karlsson, C. Kelleher, R. Kirkpatrick, M. Kirst, A. Kohler, U. Kalluri, F. Larimer, J. Leebens-Mack, J.C. Leple, P. Locascio, Y. Lou, S. Lucas, F. Martin, B. Montanini, C. Napoli, D.R. Nelson, C. Nelson, K. Nieminen, O. Nilsson, V. Pereda, G. Peter, R. Philippe, G. Pilate, A. Poliakov, J. Razumovskaya, P. Richardson, C. Rinaldi, K. Ritland, P. Rouze, D. Ryaboy, J. Schmutz, J. Schrader, B. Segerman, H. Shin, A. Siddiqui, F. Sterky, A. Terry, C.J. Tsai, E. Uberbacher, P. Unneberg, J. Vahala, K. Wall, S. Wessler, G. Yang, T. Yin, C. Douglas, M. Marra, G. Sandberg, Y. Van de Peer, and D. Rokhsar. 2006. The genome of black cottonwood, Populus trichocarpa (Torr. & Gray). *Science* 313:1596–1604.

United Nations. 1999. The world at six billion. Accessed on November 8, 2007 at *http://www.un.org/esa/population/publications/sixbillion/sixbillion.htm.*

University of Arizona. 2007. Teacher Internship in Plant Genomics Summer 2006. Accessed on September 28, 2007 at *http://biotech.biology.arizona.edu/RET/TIPG.html.*

University of Arkansas. 2007. Workshops Organized by RiceCAP. Accessed on August 22, 2007 at *http://www.uark.edu/ua/ricecap/outreach_workshops.htm.*

University of California, Davis. 2007. Wheat CAP: Coordinated Agricultural Project. Accessed on August 22, 2007 at *http://maswheat.ucdavis.edu.*

Vasil, I.K. 2003. The science and politics of plant biotechnology—a personal perspective. *Nature Biotechnology* 21:849–851.

Virginia Polytechnic Institute and State University. 2006. Fralin Biotechnology Center: Other Outreach Programs. Accessed on August 12, 2007 at *http://www.biotech.vt.edu/outreach/programs.html.*

Washington University. 2005. Genomics in Education. Accessed on August 23, 2007 at *http://www.nslc.wustl.edu/elgin/genomics/index.html.*

Whitham, T.G., J.K. Bailey, J.A. Schweitzer, S.M. Shuster, R.K. Bangert, C.J. LeRoy, E.V. Lonsdorf, G.J. Allan, S.P. DiFazio, B.M. Potts, D.G. Fischer, C.A. Gehring, R.L. Lindroth, J.C. Marks, S.C. Hart, G.M. Wimp, and S.C. Wooley. 2006. A framework for community and ecosystem genetics: from genes to ecosystems. *Nature Reviews Genetics* 7:510–523.

Wiley, H.S., S.Y. Shvartsman, and D.A. Lauffenburger. 2003. Computational modeling of the EGF-receptor system: a paradigm for systems biology. *Trends in Cell Biology* 13:43–50.

Wood, W.B., and J.M. Gentile. 2003. Teaching in a research context. *Science* 302:1510.

Yan, L., A. Loukoianov, G. Tranquilli, M. Helguera, T. Fahima, and J. Dubcovsky. 2003. Positional cloning of the wheat vernalization gene VRN1. *Proceedings of the National Academy of Sciences USA* 100:6263–6268.

Yan, L., A. Loukoianov, A. Blechl, G. Tranquilli, W. Ramakrishna, P. SanMiguel, J.L. Bennetzen, V. Echenique, and J. Dubcovsky. 2004. The wheat VRN2 gene is a flowering repressor down-regulated by vernalization. *Science* 303:1640–1644.

Yan, L., D. Fu, C. Li, A. Blechl, G. Tranquilli, M. Bonafede, A. Sanchez, M. Valarik, S. Yasuda, and J. Dubcovsky. 2006. The wheat and barley vernalization gene VRN3 is an orthologue of FT. *Proceedings of the National Academy of Sciences USA* 103:19581–19586.

Yu, J., and E.S. Buckler. 2006. Genetic association mapping and genome organization of maize. *Current Opinion in Biotechnology* 17:155–160.

Yu, W., F. Han, Z. Gao, J.M. Vega, and J.A. Birchler. 2007. Construction and behavior of engineered minichromosomes in maize. *Proceedings of the National Academy of Sciences USA* 104:8924–8929.

Zauhar, R.J. 2001. University bioinformatics programs on the rise. *Nature Biotechnology* 19:285–286.

Appendixes

A

Committee Biographies

Jeffery L. Dangl *(chair)* is the John N. Couch Distinguished Professor at the University of North Carolina. He holds a Ph.D. in genetics, M.S. in biological sciences, and B.A.S. in biological sciences and English, all from Stanford University. Dr. Dangl's research topics include using *Arabidopsis* to identify and isolate loci necessary for disease resistance responses, molecular mechanisms of disease resistance and plant cell death, *Pseudomonas syringae* Type III effectors, bacterial and plant genomics, and ancient immune systems. Dr. Dangl received the John L. Sanders Award for Distinguished Undergraduate Teaching/Service from the University of North Carolina, Chapel Hill, the Prize for Young Researchers from the State of Nord-Rhein-Westfalen in Germany in 1991, and a Plant Molecular Biology Post-Doctoral Fellowship from the National Science Foundation in 1986–1989. He chaired the National Research Council (NRC) Committee on the Plant Genome Research Initiative 2003–2008 and is a member of the NRC Board on Life Sciences. He currently serves as a member of the Sainsbury Laboratory Council and the Two Blades Foundation scientific advisory board. He was elected to the German Academy of Natural Sciences ("Die Leopoldina") in 2003 and the National Academy of Sciences in 2007.

Lois Banta is an associate professor of biology at Williams College. She received her B.A. from Johns Hopkins University and her Ph.D. from the California Institute of Technology. She studies the interactions between the soil pathogen Agrobacterium tumefaciens and its host plants. In particular, she is interested in the mechanism by which the bacterium delivers a large fragment of DNA into the host plant's genome, which is the basis for plant genetic engineering, and in the defenses mounted by the plant in response to this pathogen. Dr. Banta has served as the honors thesis adviser to 45 students and has mentored summer/winter projects for more than 60 students. She also teaches courses in agricultural biotechnology in developing economies and in bioinformations, genomics, and proteomics at Williams College. Dr. Banta served as a thesis committee member for two Ph.D. students at the University of Pennsylvania Plant Science Institute. As a Fulbright Senior Fellow in 2000, she taught and carried out research at the Institute for Molecular Plant Sciences at Leiden University in the Netherlands. Over the past 14 years, as a liberal arts college faculty member, she has received continual grant support from the National Science Foundation and has served on NSF review panels on metabolic biochemistry, signal transduction, prokaryotic molecular and cellular biology, and symbiosis, defense, and self-recognition.

Roger Boerma is a distinguished professor in the Crop and Soil Sciences Department at the University of Georgia. He received his B.S. in general agriculture and agriculture education from Illinois State University and his M.S. and Ph.D. in plant breeding and genetics from the University of Illinois. His responsibilities at the University of Georgia include development of superior-yielding, multiple pest-resistant soybean cultivars; development of molecular technologies to improve the efficiency of soybean cultivar development; identification and characterization of useful genetic variation for soybean improvement; training of graduate students and postdoctoral fellows in molecular breeding and genetics; teaching a graduate course entitled Quantitative Aspects of Molecular Breeding; program implementation for the Center for Soybean Improvement; and coordination of academic research for the Center for Applied Genetic Technologies. He was responsible for the creation of the University of Georgia Center for Soybean Improvement and currently serves as its director. Dr. Boerma also participated in the development of the University of Georgia Center for Applied Genetic Technologies Initiative (CAGT) and is currently serving as its director for academic plant research. He was the recipient of the American Soybean Association Soybean Production Research Award, American Society of Agronomy Agronomic Achievement Award, Illinois State University Alumni Achievement Award, National Council of Commercial Plant Breeders Genetics and Plant Breeding Award, Crop Science of America Research Award, and Southern Soybean Disease Workers Distinguished Service Award. He is a fellow of the Crop Science Society of America, American Society of Agronomy, and American Association for the Advancement of Science.

James C. Carrington is the director of The Center for Genome Research and Biocomputing and professor of botany and plant pathology at Oregon State University. He holds a Ph.D. from the University of California, Berkeley and a B.S. from the University of California, Riverside. Dr. Carrington's research includes small RNA pathways in plants, particularly mechanisms of microRNA-mediated suppression of targets and the roles of microRNA-guided processes during development, and also virus-host interactions controlling compatibility and defense by investigating adaptive antiviral host responses, namely RNA silencing. Dr. Carrington is a recipient of the American Phytopathological Society's Ruth Allen Award and has served on the Committee on Genetically Modified Pest Protected Plants.

Joanne Chory is a professor in the Plant Biology Laboratory at the Salk Institute for Biological Studies and an adjunct professor of biology at the University of California, San Diego. She holds a Ph.D. in microbiology from the University of Illinois at Urbana-Champaign and an A.B. in biology from Oberlin College. Dr. Chory's research uncovers the signaling pathways plants use to detect changes in the sunlight that they receive. At the Salk Institute, she has recently identified a gene, *pft1*, and characterized a signaling pathway that explains how some plants adjust their growth and flowering to shade. The findings have potentially great significance for commercial farmers, who typically plant their crops so close together that the plants must compete for light. With an understanding of pft1's role, it may be possible to develop plants that still flower and set seed, even under shady conditions. Dr. Chory is among the world leaders in the study of plant responses to light. She identified many of the genes responsible for the control of plant photomorphogenesis and discovered that this process is regulated by a steroid hormone. She also cloned the receptor for this class of hormones. She is the recipient of the Initiatives in Research Award, the Charles Albert Schull Award, the Kumho Award in Plant Molecular Biology, and Scientific American's 2003 Research Leader in Agriculture. She was elected to the National Academy of Sciences in 1999.

Steve A. Kay is the dean of biological sciences and the Richard C. Atkinson chair in biological sciences at the University of California, San Diego. He was the chair of the Department of Biochemistry at The Scripps Research Institute from 1996–2007. Prior to joining Scripps, he founded the Phenomix Corporation and served as director of discovery research at the Genomics Institute of the Novartis Research Foundation. Dr. Kay received his Ph.D. and B.S. in biochemistry at the University of Bristol in the United Kingdom. His research focuses on genetics and genomics of circadian clocks by studying the construction and evolution of complex genetic networks that underlie circadian rhythms in animals and plants through the development and use of cutting-edge technologies for measuring transcription in live cells, tissues, and intact organisms. He is the recipient of the Honma Prize for Life Sciences, the W.M. Keck Foundation Junior Faculty Award, and the NIH Director's Shannon Award. He holds three patents and currently serves on the board of the Center of Gene Research and Biotechnology and as member of the Sainsbury Laboratory.

Suzanna Lewis is the leader of the informatics group in the Berkeley *Drosophila* Genome Project at the Lawrence Berkeley Laboratory (LBL) and a founder of the Gene Ontology Consortium. She was a scientist with the Human Genome Center's informatics group that developed the software to address problems related to the electronic capture, representation, and organization of data generated by the genome laboratories. The group works with biologists and engineers at the center to provide many forms of computer assistance, including custom software, access to external databases, and tools for portability of data. In addition, the group is concerned with longer-range problems of sequence and clone assembly, database design, and tools for data management.

Thomas Mitchell-Olds is professor of biology at Duke University. He obtained his Ph.D. in botany and in the Program in Plant Breeding and Plant Genetics at the University of Wisconsin and did his postdoctoral studies in human genetics. He spent nine years as the director of the Max-Planck Institute of Chemical Ecology and has recently returned to the United States. He uses an interdisciplinary approach incorporating functional genomics, physiology, and population and quantitative genetics to understand the genetic architecture and evolutionary significance of complex trait variation within and among plant populations. His ongoing experiments focus on wild relatives of *Arabidopsis*, as well as an expanding emphasis on drought tolerance in cereals and comparative genomics of grasses.

Neelima R. Sinha is a professor of plant biology at the University of California, Davis. Dr. Sinha holds a Ph.D. in Botany from the University of California, Berkeley and an M.S. in environmental studies from Baylor University. Her research focuses on studying fundamental mechanisms of leaf development using model organisms, such as tomato and *Arabidopsis*. To understand how leaves evolved, she and members of her laboratory are studying other organisms like *Welwitschia mirabilis*, *Kalanchoe diagremontinan*, ferns, cycads, and basal and derived angiosperms. Her collaborative research with Kenyatta University in Nairobi aims at improving maize resistance to the parasitic weed *Striga hermonthica*. She is the recipient of the Katherine Esau Junior Faculty Fellowship and the Chancellor's Award for Excellence in Undergraduate Research Mentorship.

Michael Snyder is the Lewis B. Cullman Professor of Molecular, Cellular and Developmental Biology and a professor of molecular biophysics and biochemistry at Yale University. He is also the director of the Yale Center for Genomics and Proteomics. Dr. Snyder holds a Ph.D. in biology from the California Institute of Technology and a B.A. in chemistry and biology from the University of Rochester. He uses global approaches to explore protein function and dissect regulatory networks. Several of his areas of research are control of cell division and cell morphogenesis in yeast, characterization of proteomes, and analysis of regulatory circuits in yeast. He was a recipient of the Burroughs Wellcome Scholar Award, the Pew Scholar Award, the Helen Hay Whitney Postdoctoral Fellowship, and the NIH Predoctoral Training Fellowship. He was a member of the NRC HHMI Predoctoral Fellowships Panel on Cell Biology and Immunology.

Steven H. Strauss is a professor in the Forest Science, Molecular and Cellular Biology, and Genetics Programs at Oregon State University. He is also director of the university's Program in Outreach in Biotechnology, which seeks to promote public and professional understanding of the potential benefits and risks associated with natural resource biotechnologies. He holds a Ph.D. in genetics from the University of California at Berkeley, an M.F.S. in forest science from Yale University, and a B.S. in biology from Cornell University. Dr. Strauss's current research focuses on modifying the architecture, chemistry, and flowering of poplars for wood, bioproduct, and energy uses. He directs a university-industry research cooperative that aims to advance knowledge of genomics and biosafety technologies for genetic engineering of trees. He is a Stanford Institute for the Environment Leopold Fellow (2005). He served on the NRC Committees to Review the R&D Strategy for Biomass-Derived Ethanol and Biodiesel Transportation Fuels, and on the Impacts of GMOs on Terrestrial and Aquatic Wildlife and Habitats. He has edited two books and published more than 150 scientific and professional publications.

Eric R. Ward is the president of the Two Blades Foundation. He received his Ph.D. in plant biology from Washington University in St. Louis, where he was a graduate fellow of the National Science Foundation. He received his B.S in biology magna cum laude from Duke University. Dr. Ward served most recently as co-president of Novartis (now Syngenta) Agribusiness Biotechnology Research, where he was responsible for a staff of 270, including researchers and all administrative functions such as finance, patents, business development, public affairs, human resources, and facilities. Simultaneously, he was head of target discovery for Novartis Crop Protection AG, where he implemented a fully integrated agricultural chemical lead discovery program based on proprietary molecular targets. This program relied on extensive interactions with biotechnology firms and academic laboratories.

Prior to that, he was a research director for the Novartis herbicide business unit, during which time his team invented the Acuron™ herbicide tolerance technology, developed corn and sugar beet varieties engineered with the Acuron™ gene, and built the patent strategy to protect the technology. Prior to that, he worked in Basel, Switzerland, as a project leader for Ciba Crop Protection in the Weed Control business unit. Dr. Ward began his career in 1988 with Ciba-Geigy as a postdoctoral associate, during which time he pioneered methods for the cloning of large DNA fragments from plants. Dr. Ward has authored more than 65 scientific publications and is inventor on more than 20 issued patents.

B

Publications in Genomics of the Top 40 Most-Cultivated Crops

The 40 most-cultivated species were found by searching the Food and Agriculture Organization of the United Nations' FAOSTAT agricultural database of primary crops by area harvested. Model species are provided as references. For each species, the Web of Science, a database of about 8,700 journals, was queried for articles with the species name or common name in the title (ti=common name or ti=species name). The common name in plural and singular forms (for example, orange and oranges) and common names that can be compound words or two words (for example, chick pea and chickpea) were all used in the search. To narrow down the list of articles to those relevant to genomics, only the articles that include either *genomics, genomic, sequencing,* or *sequence* as a topic are included (tc=genome or tc=genomics or tc=sequencing or tc=sequence).

Most Cultivated Crops in 2005	Rank in 2005	Rank in 2000	Area Cultivated in 2005 (HA)	Area Cultivated in 2002 (HA)
Wheat	1	1	214,666,735	215,474,117
Rice, paddy	2	2	152,904,158	153,942,855
Maize	3	3	143,798,427	139,494,325
Soybeans	4	4	92,443,356	74,321,496
Barley	5	5	55,895,782	54,489,630
Sorghum	6	6	43,360,759	40,831,541
Seed cotton	7	8	35,003,800	31,924,034
Millet	8	7	34,815,044	37,006,519
Rapeseed	9	9	27,448,195	25,833,433
Beans, dry	10	10	25,231,390	23,531,811
Groundnuts, with shell	11	11	23,469,008	23,358,847
Sunflower seed	12	12	22,372,946	21,161,673
Potatoes	13	13	18,872,538	20,130,298
Sugar cane	14	14	18,730,869	19,466,758
Cassava	15	15	18,320,052	17,059,381
Alfalfa for forage and silage	16	16	15,118,543	15,421,302
Oats	17	17	11,450,461	12,676,446
Coconuts	18	18	10,681,368	10,596,868
Chick peas	19	20	10,646,032	10,100,570
Coffee, green	20	19	10,007,069	10,317,198
Oil palm fruit	21	21	9,384,653	9,954,304
Cow peas, dry	22	25	9,170,820	7,494,030
Sweet potatoes	23	23	8,810,860	9,703,499
Natural rubber	24	31	8,114,849	5,640,296
Sesame seed	25	27	7,596,056	7,269,734
Olives	26	24	7,574,361	8,327,400
Grapes	27	26	7,470,679	7,377,007
Rye	28	22	6,829,860	9,818,138
Peas, dry	29	30	6,298,663	5,769,319
Rye grass for forage and silage[a]	30		5,505,313	
Plantains	31	33	5,399,072	4,976,685
Cocoa beans	32	28	5,287,847	7,125,153
Sugar beet	33	29	5,240,711	6,011,351
Apples	34	32	4,774,376	5,310,185
Pigeon peas	35	34	4,572,039	4,244,022
Tomatoes	36	38	4,570,870	3,968,508
Bananas	37	37	4,370,342	4,030,079
Lentils	38	39	4,084,614	3,873,738
Tobacco	39	35	3,908,731	4,160,103
Triticale[a]	40		3,744,001	
Model species				
Arabidopsis				
Brachypodium				
Chlamydomonas				
Lotus				
Populus				
Medicago				

[a]These crops were not one of the top 40 most cultivated crops in 2001.

Total Number of Publications 2005-2006	Total Number of Publications 2000-2001	Total Number of Institutions Contributing to the Papers 2005-2006	Total Number of Institutions Contributing to the Papers 2000-2001	Percent Publications That Involve U.S. Institutions 2005	Percent Publications That Involve U.S. Institutions 2000
473	357	434	305	28	20
636	365	561	364	22	23
308	232	337	222	48	54
35	152	63	171	40	55
172	131	213	173	26	25
55	32	88	44	53	50
45	12	48	12	51	75
35	24	55	31	31	38
152	56	187	79	14	14
37	18	71	28	24	22
19	16	40	22	37	44
14	12	19	17	43	25
204	182	263	202	21	18
7	2	14	4	14	0
19	25	32	34	37	20
33	38	46	33	45	39
18	36	24	40	28	44
6	7	14	12	17	0
28	13	36	16	14	31
31	10	50	23	19	20
5	5	16	7	0	0
16	34	28	20	13	29
24	18	33	31	8	11
12	12	30	17	58	25
4	4	8	10	25	0
56	33	77	40	5	0
48	24	65	37	13	33
27	38	44	58	11	18
21	23	32	36	14	17
35		57		6	
23	18	46	25	22	17
4	3	7	5	25	0
23	22	35	26	22	9
63	48	102	69	17	38
4	3	8	3	0	0
184	159	275	176	34	37
5	33	8	37	140	18
53	7	80		0	
185	143	246	254	2	43
13		15		23	
781	475	642	403	41	40
4	0	9	0	50	0
67	57	95	74	48	33
33	22	69	39	9	32
47	19	80	30	36	32
51	21	78	34	41	0

Most Cultivated Crops in 2005	Rank in 2005	Rank in 2000	Times That the Most Frequently Cited Paper Has Been Cited Since Its Publication 2005-2006	Times That the Most Frequently Cited Paper Has Been Cited Since Its Publication 2000-2001
Wheat	1	1	40	175
Rice, paddy	2	2	234	233
Maize	3	3	67	152
Soybeans	4	4	14	116
Barley	5	5	33	187
Sorghum	6	6	40	65
Seed cotton	7	8	7	42
Millet	8	7	8	25
Rapeseed	9	9	23	132
Beans, dry	10	10	10	47
Groundnuts, with shell	11	11	9	30
Sunflower seed	12	12	5	31
Potatoes	13	13	26	94
Sugar cane	14	14	3	1
Cassava	15	15	10	67
Alfalfa for forage and silage	16	16	11	45
Oats	17	17	11	43
Coconuts	18	18	2	3
Chick peas	19	20	10	53
Coffee, green	20	19	8	959
Oil palm fruit	21	21	5	38
Cow peas, dry	22	25	4	66
Sweet potatoes	23	23	5	34
Natural rubber	24	31	6	34
Sesame seed	25	27	5	30
Olives	26	24	7	59
Grapes	27	26	10	101
Rye	28	22	15	28
Peas, dry	29	30	6	74
Rye grass for forage and silage[a]	30		15	
Plantains	31	33	5	23
Cocoa beans	32	28	6	36
Sugar beet	33	29	7	101
Apples	34	32	17	48
Pigeon peas	35	34	1	5
Tomatoes	36	38	36	277
Bananas	37	37	11	35
Lentils	38	39	3	10
Tobacco	39	35	41	150
Triticale[a]	40		4	
Model species				
Arabidopsis			158	385
Brachypodium			4	N/A
Chlamydomonas			25	71
Lotus			39	49
Populus			37	70
Medicago			42	66

[a]These crops were not one of the top 40 most cultivated crops in 2001.

Average Times Cited of the Top 10 Cited Papers 2005-2006	Average Times Cited of the Top 10 Cited Papers 2000-2001	Number of Top 10 Cited Papers That Involve U.S. Institutions 2005-2006	Number of Top 10 Cited Papers That Involve U.S. Institutions 2000-2001	Number of Top 10 Cited Papers That Involve U.S. and International Collaboration 2005-2006	Number of Top 10 Cited Papers That Involve U.S. and International Collaboration 2000-2001
21	100	4	2	4	1
69	137	7	7	5	1
25	95	9	10	2	2
6	59	8	4	0	1
20	97	6	3	4	2
14	35	10	7	0	1
14	21	8	9	2	4
4	14	7	4	2	1
12	46	5	2	4	0
4	20	2	2	2	1
2	14	4	5	2	1
2	11	6	2	0	0
11	68	3	1	2	0
N/A	N/A	1	0	0	0
5	32	1	3	1	3
6	30	4	2	1	0
3	24	4	8	0	1
N/A	N/A	1	0	1	0
5	21	3	3	2	2
4	114	3	2	2	1
N/A	N/A	0	0	0	0
1	30	0	2	0	0
2	16	1	0	0	0
2	13	5	1	5	0
N/A	N/A	1	0	0	0
5	29	2	0	2	0
7	32	1	4	4	4
4	20	2	1	2	1
3	34	2	2	1	2
4		2		2	
2	13	3	2	2	2
N/A	N/A	1	0	1	0
3	26	3	1	1	1
9	33	1	5	0	2
N/A	N/A	0	0	0	0
17	103	5	8	1	2
4	25	2	2	2	2
N/A	N/A	0	0	0	0
18	99	4	6	1	1
1		3		0	
64	260	7	6	2	0
N/A	N/A	2	N/A	0	N/A
9	43	9	3	4	0
11	35	2	4	2	2
15	39	4	4	2	2
16	38	4	3	2	1

List of Keywords Used in the Search for Publications

Crops	Keywords Used
Wheat	ti=wheat, ti=triticum
Rice, paddy	ti=rice, ti=oryza
Maize	ti=corn, ti=maize, ti=zea mays
Soybeans	ti=soybean, ti=soybeans, ti=Glycine max
Barley	ti=barley, ti=Hordeum vulgare
Sorghum	ti=Sorghum
Seed cotton	ti=seed cotton, ti=Gossypium
Millet	ti=millet, ti=Pennisetum, ti=Echinochloa, ti=Panicum, ti=setaria, ti=Paspalum
Rapeseed	ti=rapeseed, ti=Brassica
Beans, dry	ti=dry bean, ti=dry beans, ti=Phaeolus
Groundnuts, with shell	ti=ground nuts, ti=ground nut, ti=groundnuts, ti=groundnutArachis hypogaeati
Sunflower seed	ti=sunflower seed, ti=sunflower seeds, ti=Helianthus annuus
Potatoes	ti=potatoes, ti=potato, ti=Solanum tuberosum
Sugar cane	ti=sugar cane, ti=Saccharum officinarum
Cassava	ti=cassava, ti=Manihot esculenta
Alfalfa for forage and silage	ti=alfalfa, ti=Medicago sativa
Oats	ti=oats, ti=oat, ti=Avena
Coconuts	ti=coconuts, ti=concout, ti=Cocos nucifera
Chick peas	ti=chick peas, ti=chick pea, ti=chickpeas, ti=chickpea, ti=Cicer arietinum
Coffee, green	ti=coffee, ti=Coffea
Oil palm fruit	ti=oil palm fruits, ti=oil palm fruit, ti=Elaeis guineensis
Cow peas, dry	ti=cow peas, ti=cow pea, ti=cowpeas, ti=cowpea, ti=Vigna
Sweet potatoes	ti=sweet potatoes, ti=sweet potato, ti=Ipomoea batatas
Natural rubber	ti=natural rubber, ti=Hevea
Sesame seed	ti=sesame seeds, ti=sesame seed, ti=Sesamum indicum
Olives	ti=olives, ti=olive, ti=Olea europeaea
Grapes	ti=grapes, ti=grape, ti=Vitis vinifera
Rye	ti=rye, ti=Secale cereale
Peas, dry	ti=dry peas, ti=dry pea, ti=Pisum sativum
Rye grass for forage and silage	ti=rye grass, ti=Lolium
Plantains	ti=plantains, ti=plantain, ti=Musa
Cocoa beans	ti=cocoa beans, ti=cocoa bean, ti=Theobroma
Sugar beet	ti=sugar beets, ti=sugar beet, ti=Beta vulgaris
Apples	ti=apples, ti=apple, ti=Malus
Pigeon peas	ti=pigeon peas, ti=pigeon pea, ti=Cajanus cajan
Tomatoes	ti=tomatoes, ti=tomato, ti=Lycopersicon
Bananas	ti=bananas, ti=banana, ti=Musa
Lentils	ti=lentils, ti=lentil, ti=Lens culinaris
Tobacco, unmanufactured	ti=tobacco, ti=Nicotiana tabacum
Triticale	ti=Triticale
Model species	
Arabidopsis	ti=arabidopsis
Brachypodium	ti=Brachypodium
Chlamydomonas	ti=Chlamydomonas
Lotus	ti=Lotus
Populus	ti=Populus
Medicago	ti=Medicago

C

Questionnaire to Lead Principal Investigators of NPGI Grants

a. Indicate the numbers of students and postdocs who have been directly or partially supported in your lab by NPGI. Please DO NOT use their names, but list them as undergrad 1, undergrad 2, grad student 1, postdoc 1, etc. For each individual, please also indicate in what sector (industry, academia, government, other) they now hold positions, if known, and whether these positions are in the U.S. or abroad.

Note: If you don't know where they are now, please answer "don't know." For those who are still in your lab, write "still in my lab."

b. In 5 short bullet points, identify the 5 most important contributions you and your group have made to the goals of NPGI using your NPGI funding.

c. In a concise bulleted list, identify how those contributions are documented (e.g. either as a publication, please cite, or a Web resource, a genome sequence or other-please describe "other").

d. List ALL your publications that cite NPGI support.

e. Identify <u>explicit</u> examples of your interaction with industry or breeders in terms of translating basic research into eventual applied agricultural value.

f. Identify how these interactions are documented.

g. List up to 5 Websites or Web resources that you use the most for your NPGI research.

D

Workshop on the National Plant Genome Initiative

Keck Center of the National Academies
July 6, 2007

Welcome Remarks
James P. Collins, Cochair of the Interagency Working Group on Plant Genomes

Purpose of the Workshop
Jeffery L. Dangl, Chair of the NRC Committee on the National Plant Genome Initiative

Panel 1: Technology
Daniel Rohksar, Joint Genome Institute
Robert Reiter, Monsanto

Panel 2: Data Management
Lincoln Stein, Cold Spring Harbor Laboratory
Gerard Manning, Salk Institute
Michael Sanderson, University of California, Davis

Panel 3: Commodities Genomics
Jeffrey Bennetzen, University of Georgia, Athens
Jeffrey Dean, University of Georgia, Athens
Erik Legg, Syngenta
Patrick Schnable, Iowa State University

Panel 4: Domestication and Population Genomics
Susan McCouch, Cornell University
Edward Buckler, Cornell University
Charles Langley, University of California, Davis
Magnus Nordborg, University of Southern California

Panel 5: Ecological Genomics
Barbara Schaal, Washington University, St. Louis
Thomas Juenger, University of Texas, Austin
John Willis, Duke University

Panel 6: Functional Genomics
Mary Lou Guerinot, Dartmouth College
Joseph Kieber, University of Carolina, Chapel Hill

Panel 7: Training, Education and Outreach
Susan Singer, Carleton College
Lois Banta, Williams College

Concurrent Breakout Sessions 1

1a) Sequenced Plant and Associated Data Management
All major crops and common forms thereof and all major pathogens
Sample sequence of all major clades

1b) Regulatory Plant
Understanding regulatory networks at all levels

1c) Evolving Plant
Understanding the evolutionary history of all major plants, their diversity, how it has changed and how it is changing

Concurrent Breakout Sessions 2

2a) Dynamic Plant
Understanding plant cell differentiation, cell-cell interaction and changes over developmental, time and in response to both biotic and abiotic stresses

2b) Community Plant
Genome by genome and genome(s) by environment interactions
2c) Mathematical Plant
Predictive power to model and test models at all levels of biological organization, from molecular machines to populations.

July 7, 2007

Concurrent Breakout Sessions 3
3a) Deployed Plant
Translation to farm and field.
3b) Educated Plant
Training, education, and outreach
3c) Agency Strategy

E

Summary of Grants Given by the National Plant Genome Initiative

TABLE E-1 Summary of Grants Given by the National Plant Genome Initiative Through Various Member Agencies

Year	Number of Grants Given	Number of Different Principal Investigators in Each Year	Total Number of Principal Investigators Funded since 1998	Total Value	Range
1998	21	21	21	$75,931,799	$301,072–13,117,560
1999	42	39	59	$82,742,366	$49,990–10,905,763
2000	40	39	91	$59,027,689	$89,712–7,600,000
2001	39	37	119	$86,315,777	$90,000–7,471,263
2002	48	46	152	$87,131,418	$73,169–8,368,574
2003	57	53	190	$99,552,047	$75,000–10,911,015
2004	40	40	218	$89,516,090	$105,000–10,261,784
2005	46	44	245	$98,747,726	$20,000–29,450,001
2006	59	56	**277**	$95,079,294	$10,000–6,590,671
Total	**392**	**375**		**$774,044,206**	

TABLE E-2 Amount of Funds Given to Institutions in Different States for Equipment Through the National Plant Genome Initiative from 1996-2006. Because the Amount for Equipment Was Not Listed for 177 Grants of a Total Of $60,285,789, This Table Only Represents the Amount of Funds from 215 Grants of a Total of $713,758,417

State	1998	1999	2000	2001
AL				
AR				
AZ		$155,463		
CA	$821,829	$207,252	$431,354	$29,000
CT			$64,700	$21,500
DE	$95,000			$4,134
FL			$86,500	
GA	$55,900	$8,800	$10,000	$374,192
IA		$54,459		$13,000
IL	$264,039			
IN		$28,000	$133,100	
KS				
MA		$131,500		
MD	$62,000	$88,800		
MI	$182,152			$40,000
MN	$305,610			$58,384
MO	$154,000	$5,749		$14,000
MS				
NC		$300,000		$43,000
NE				
NH			$7,122	
NJ	$5,000	$443,222	$80,085	
NV				
NY	$198,872		$29,900	$171,769
PA	$232,015			$87,575
OH				$154,273
OK				
OR			$26,000	
SC		$26,000		
SD				
TX				
UT				$22,612
VA				$51,684
WA			$949,778	
WI				
Total by year	$2,382,175	$1,449,245	$1,818,539	$1,085,123
Total	**$13,668,174**			

2002	2003	2004	2005	2006
$137,925				
		$11,500		
$13,750	$497,250	$6,000	$33,563	
$131,000	$241,500			$5,001
$376,000	$444,475	$25,000		
	$11,000			
$9,500	$156,745		$100,000	$23,500
		$111,000		
$36,000	$82,340			$223,274
	$39,000	$73,000	$47,000	
$393,100			$145,000	$69,707
			$310,000	
	$13,339			$103,565
$58,288			$29,250	$15,700
$60,000	$24,363			$17,000
$209,500	$200,000	$120,000		
		$101,000		
	$34,000	$347,000		$18,500
$27,480				
				$47,942
	$368,450			
$114,000	$48,000			
$65,000	$648,216	$170,000	$15,764	$32,240
		$17,800	$10,000	
$45,000				
				$50,357
				$47,000
				$17,000
	$27,500			
$14,034				
			$23,825	$20,849
			$18,000	
$1,690,577	$2,836,178	$982,300	$732,402	$691,635

TABLE E-3 List of Institutions with Which the Lead Principal Investigators of NPGI Grants Are Affiliated

Institution	Number of Different Lead Principal Investigators
Agricultural Research Service	
Midwest Area	5
North Atlantic Area	2
Northern Plains Area	2
Pacific West Area	5
South Atlantic Area	2
Alabama A&M University	1
Brookhaven National Laboratory	1
California State Polytechnic University	1
Carnegie Institute of Washington	1
Clemson University	6
Cold Spring Harbor Laboratory	5
Colorado State University	1
Cornell University	13
Dartmouth College	3
Donald Danforth Plant Science Center	2
Duke University	1
F. Hutchinson Cancer Research Center	1
Florida State Unviersity	1
Forest Service, Pacific Southwest Research Station	1
Harvard University	1
Indiana University, Bloomington	4
Institute for Genomic Research	3
Institute of Paper Science and Technology	1
Iowa State University	9
Kansas State University	4
Lawrence Berkeley National Laboratory	1
Louisiana State University & A&M College	3
Massachusetts Institute of Technology	1
Michigan State University	9
Michigan Technological University	2
Mississippi State University	2
New York University	1
North Carolina State University	8
North Dakota State University	3
Mayville State University	1
Oak Ridge National Laboratory	2
Ohio State University	4
Oregon State University	2
Pennsylvania State University	4
Portland State University	1
Purdue University	6
Rice University	1
Rutgers University	3
Salk Institute	1
South Dakota State University	1
Southern Illinois University, Carbondale	2

TABLE E-3 Continued

Institution	Number of Different Lead Principal Investigators
Stanford University	1
State University of New York, Stony Brook	1
Tennessee State University	1
Texas A&M University	3
Texas Tech University	1
Samuel Roberts Noble Foundation	2
University of Alabama, Birmingham	1
University of Arizona	8
University of Arkansas, Fayetteville	2
University of California, Berkeley	6
University of California, Davis	16
University of California, Los Angeles	2
University of California, Riverside	4
University of California, San Diego	1
University of Chicago	1
University of Connecticut	1
University of Delaware	3
University of Florida	5
University of Georgia	12
University of Illinois, Urbana-Champaign	8
University of Kentucky	4
University of Maine, Orono	1
University of Minnesota, St. Paul	3
University of Minnesota, Twin Cities	9
University of Missouri, Columbia	9
University of Nebraska, Lincoln	6
University of Nevada, Las Vegas	1
University of Nevada, Reno	3
University of New Hampshire	1
University of North Carolina, Chapel Hill	1
University of North Carolina, Wilmington	1
University of North Dakota	1
University of Northern Colorado	1
University of Oregon	1
University of Pittsburgh	1
University South Carolina	1
University of South Dakota	1
University of Texas, Austin	1
University of Utah	1
University of Vermont	1
University of Washington	2
University of Wisconsin, Madison	7
Vanderbilt University	1
Virginia Polytechnic Institute and State University	3
Washington State University	3
Washington University	2
Yale University	4
Total number of institutions = 89	

F

List of Websites That Lead Principal Investigators of NPGI Grants Reported as the Top Five Websites Used for NPGI Research

Website	Number of PIs Who Reported It as One of the Top Five Websites
http://www.ncbi.nlm.nih.gov/	120
http://www.arabidopsis.org/[a]	56
http://www.tigr.org	51
http://www.gramene.org [a]	37
http://www.maizegdb.org[a]	26
http://wheat.pw.usda.gov/GG2/index.shtml[a]	21
http://compbio.dfci.harvard.edu/tgi/	12
http://www.plantgdb.org	10
http://www.jgi.doe.gov/	9
http://signal.salk.edu	8
http://www.bioinfo.wsu.edu/gdr[a]	7
http://www.genome.arizona.edu	7
http://www.google.com/	7
http://www.maizesequence.org	7
http://expasy.org/	6
http://www.ars-grin.gov/	6
http://www.chromdb.org	6
http://www.comparative-legumes.org[a]	6
http://www.plantgenomics.iastate.edu/maize/	6
http://www.plexdb.org/	6
http://www.sgn.cornell.edu/[a]	6
IISI Web of Knowledge	6
http://mpss.udel.edu/	5
http://www.ebi.ac.uk/	5
http://www.medicago.org/	5
http://www.softberry.com/berry.phtml	5
http://magi.plantgenomics.iastate.edu	4
http://soybase.agron.iastate.edu/	4
http://compgenomics.ucdavis.edu/	3
http://dendrome.ucdavis.edu/[a]	3
http://frodo.wi.mit.edu/cgi-bin/primer3/primer3_www.cgi	3
http://maswheat.ucdavis.edu/	3
http://mips.gsf.de/	3
http://www.biosci.ohio-state.edu/pcmb/Facilities/abrc/abrchome.htm	3
http://www.cazy.org/	3

Website	Number of PIs Who Reported It as One of the Top Five Websites
http://www.geneontology.org/	3
http://www.genome.clemson.edu	3
http://www.panzea.org/	3
https://www.genevestigator.ethz.ch/	3
http://asrp.cgrb.oregonstate.edu	2
http://bioinformatics.med.yale.edu/rc/overview.jspx	2
http://cggc.agtec.uga.edu/	2
http://cgpdb.ucdavis.edu/	2
http://harvest.ucr.edu/	2
http://microrna.sanger.ac.uk/	2
http://rgp.dna.affrc.go.jp	2
http://virtualplant.org	2
http://www.agcol.arizona.edu/pave/cotton/	2
http://www.bar.utoronto.ca/	2
http://www.broad.mit.edu/annotation/fgi/	2
http://www.ccrc.uga.edu/~mao/cellwall/main.htm	2
http://www.genome.jp/kegg/	2
http://irri.org/	2
http://www.maizearray.org/	2
http://www.phytozome.net/sorghum	2
http://www.plantgenome.uga.edu	2
http://www.pseudomonas-syringae.org/	2
http://www.sciencemag.org/	2
http://www.shigen.nig.ac.jp/rice/oryzabase/top/top.jsp	2
http://www.tomatomap.net/	2
http://www.weigelworld.org/resources/microarray/AtGenExpress/	2
http://affymetrix.arabidopsis.info/narrays/experimentbrowse.pl	1
http://agricola.nal.usda.gov/	1
http://aztec.stanford.edu/gfp/	1
http://barleycap.coafes.umn.edu/[a]	1
http://barleypop.vrac.iastate.edu/BarleyBase/BarleyBase/	1
http://bioweb.pasteur.fr/intro-uk.html	1
http://brassica.bbsrc.ac.uk/	1
http://cabdirect.org/	1
http://cdna01.dna.affrc.go.jp/cDNA/	1
http://cellwall.genomics.purdue.edu	1

Website	Number of PIs Who Reported It as One of the Top Five Websites
http://charge.ucdavis.edu/	1
http://comparative-legumes.org/	1
http://compositdb.ucdavis.edu/	1
http://cottonevolution.info/	1
http://depts.washington.edu/agro/	1
http://diurnal.cgrb.oregonstate.edu/	1
http://en.wikipedia.org/wiki/Main_Page	1
http://evolution.genetics.washington.edu/phylip.html	1
http://frontend.bioinfo.rpi.edu/applications/mfold/cgi-bin/rna-form1.cgi	1
http://fungen.org/genediscovery/	1
http://genetics.mgh.harvard.edu/sheenweb/main_page.html	1
http://java.sun.com/j2se/1.5.0/docs/api/allclasses-noframe.html	1
http://jsp.weigelworld.org	1
http://linkage.rockefeller.edu/soft/mapmaker/	1
http://maize-mapping.plantgenomics.iastate.edu/	1
http://niblrrs.ucdavis.edu/	1
http://nybg.bio.nyu.edu/vicogenta	1
http://nypg.bio.nyu.edu/orthologid	1
http://orygenesdb.cirad.fr	1
http://pfam.janelia.org	1
http://phymap.ucdavis.edu:8080/barley/index.jsp	1
http://pir.georgetown.edu	1
http://plantgenomics.biology.yale.edu/	1
http://plantrbp.uoregon.edu/	1
http://plantsp.genomics.purdue.edu/	1
http://popgenome.ag.utk.edu/mdb	1
http://psbc.ucdavis.edu/home/	1
http://rana.lbl.gov/EisenSoftware.htm	1
http://ricegaas.dna.affrc.go.jp/usr/	1
http://rkd.ucdavis.edu/	1
http://searchlauncher.bcm.tmc.edu/	1
http://sorgblast3.tamu.edu/index.html	1
http://soybeangenome.siu.edu	1
http://statgen.ncsu.edu/qtlcart/	1
http://string.embl.de/	1
http://ted.bti.cornell.edu/	1

Website	Number of PIs Who Reported It as One of the Top Five Websites
http://tilling.fhcrc.org:9366	1
http://titan.biotec.uiuc.edu/apple/background.shtml	1
http://tomet.bti.cornell.edu/	1
http://tools.neb.com/NEBcutter2/index.php	1
http://tos.nias.affrc.go.jp	1
http://users.unimi.it/hidras/	1
http://wheatdb.ucdavis.edu:8080/wheatdb/index.jsp	1
http://workbench.sdsc.edu	1
http://www.affymetrix.com/analysis/index.affx	1
http://www.aphis.usda.gov/	1
http://www.arabidopsisthaliana.com/	1
http://www.avrdc.org/	1
http://www.bio.purdue.edu/people/faculty/gelvin/gelvinweb/main.html	1
http://www.bioinformatics.org/	1
http://www.bioinformatics2.wsu.edu/cgi-bin/Athena/cgi/home.pl	1
http://www.blast2go.de	1
http://www.bmb.leeds.ac.uk/staff/jpk/antibodies.htm	1
http://www.brassica.info/[a]	1
http://www.ccrc.uga.edu/~mao/cellwall/main.htm	1
http://www.ccts.uky.edu/Research/microRNA/miRNADesignOne.aspx	1
http://www.conifergdb.org/	1
http://www.faculty.ucr.edu/~tgirke/Cellwall/index.html	1
http://www.facultyof1000.com/search/pm/advanced	1
http://www.fgsc.net	1
http://www.fungen.org/	1
http://www.genome.ou.edu/medicago.html	1
http://www.gmod.org/	1
http://www.graingenes.org[a]	1
http://www.libraries.rutgers.edu/rul/rr_gateway/ejournals/ejournals.shtml	1
http://www.maizegenetics.net/	1
http://www.mgel.msstate.edu/	1
http://www.mgosdb.org/	1
http://www.mimulusevolution.org/	1
http://www.nature.com/index.html	1
http://www.ncgr.org/ourwork/#lis	1
http://www.nhm.ac.uk/research-curation/projects/solanaceaesource//	1

Website	Number of PIs Who Reported It as One of the Top Five Websites
http://www.noble.org/medicago	1
http://www.nsf.gov/	1
http://www.oardc.ohio-state.edu/vanderknaap	1
http://www.omap.org	1
http://www.ornl.gov/sci/ipgc/ [a]	1
http://www.phytome.org	1
http://www.pir.uniprot.org/	1
http://www.plantgenomeoutreach.eeob.iastate.edu/index.htm	1
http://www.pnas.org	1
http://www.premierbiosoft.com/netprimer/netprlaunch/netprlaunch.html	1
http://www.proweb.org/input/	1
http://www.psc.edu/biomed/genedoc	1
http://www.purdue.edu/discoverypark/ionomics	1
http://www.repeatmasker.org	1
http://www.sanger.ac.uk/	1
http://www.scabusa.org/	1
http://www.scholar.google.com	1
http://www.sciencemuseum.org.uk/antenna/waterproofrice/	1
http://www.tm4.org/mev.html	1
http://www.uniprot.org	1
http://www.wikipedia.org/	1
http://www.zincfingers.org	1
http://xyloglucan.prl.msu.edu	1
https://www.fastlane.nsf.gov/fastlane.jsp	1
KIREN	1
LIS	1
VectorNTI	1

[a]Denotes Websites that are listed in Table 2-3 as examples of Websites that are direct results of the National Plant Genome Initiative and the Interagency Working Group on Plant Genomes' collaborations.

G

Impact Factor of Journals in Which Awardees of NPGI Grants Published Their Articles

Journal Title	Articles	Impact Factor in 2006	Cumulative Number of Publications
Science	20	30.028	20
Cell	3	29.194	23
Nature	19	26.681	42
Nature Genetics	9	24.176	51
Nature Reviews Genetics	3	22.947	54
Nature Biotechnology	6	22.672	60
Annual Review of Plant Biology	8	19.837	68
Annual Review of Genetics	1	19.098	69
Microbiology and Molecular Biology Reviews	1	15.864	70
Nature Reviews Microbiology	2	15.845	72
Genes and Development	2	15.05	74
Nature Methods	1	14.959	75
Annual Review of Microbiology	2	14.553	77
PLoS Biology	2	14.101	79
Developmental Cell	1	13.523	80
Trends in Cell Biology	1	12.429	81
Current Biology	1	10.988	82
Genome Research	53	10.256	135
Angewandte Chemie - International Edition	1	10.232	136
Current Opinion in Plant Biology	27	10.182	163
EMBO Journal	4	10.086	167
Current Opinion in Genetics and Development	2	10.006	169
Trends in Genetics	10	9.95	179
Plant Cell	64	9.868	243
Proceedings of the National Academy of Sciences USA	74	9.643	317
Molecular and Cellular Proteomics	1	9.62	318
Annual Review of Phytopathology	3	9	321
Seminars in Cell and Developmental Biology.	1	8.672	322
Trends in Microbiology	3	8.335	325
Trends in Plant Science	11	8	336
Trends in Biotechnology	1	7.843	337
Development	5	7.764	342
Journal of American Chemical Society	1	7.696	343
PLoS Genetics	6	7.671	349
Current Opinion in Microbiology	1	7.445	350
Genome Biology	21	7.172	371
Current Opinion in Biotechnology	4	6.949	375
Molecular Biology and Evolution	13	6.726	388

Journal Title	Articles	Impact Factor in 2006	Cumulative Number of Publications
Plant Journal	73	6.565	461
Molecular Biology of the Cell	3	6.562	464
Journal of Cell Science	2	6.427	466
Nucleic Acids Research	42	6.317	508
Plant Physiology	127	6.125	635
PLoS Pathogens	2	6.056	637
BioEssays	3	5.965	640
Journal of Biological Chemistry	5	5.808	645
Proteomics	2	5.735	647
Molecular Microbiology	3	5.634	650
BioSciences	1	5.424	651
Journal of Virology	1	5.341	652
Journal of Proteome Research	2	5.151	654
RNA	2	5.111	656
Cell Microbiology	2	5.07	658
PLoS Computational Biology	1	4.914	659
Bioinformatics	36	4.894	695
Developmental Biology	3	4.893	698
Journal of Molecular Biology	2	4.89	700
Molecular Ecology	7	4.825	707
Ecology	1	4.782	708
Philosophical Transactions of the Royal Society London B,	3	4.579	711
BMC Evolutionary Biology	2	4.455	713
Evolution	2	4.292	715
New Phytologist	14	4.245	729
Genetics	134	4.242	863
Plant Cell and Environment	3	4.135	866
Electrophoresis	2	4.101	868
Chromosoma	3	4.065	871
BMC Genomics	19	4.029	890
Journal of Bacteriology	7	3.993	897
Molecular Plant-Microbe Interactions	32	3.936	929
Eukaryotic Cell	1	3.707	930
Journal of Experimental Botany	11	3.63	941
BMC bioinformatics	6	3.617	947
Plant Molecular Biology	45	3.577	9924
Genomics	5	3.558	997
Journal of Chromatography A	1	3.554	998

Journal Title	Articles	Impact Factor in 2006	Cumulative Number of Publications
Applied and Environmental Microbiology	3	3.532	1001
Molecular Phylogenetics and Evolution,	2	3.528	1003
BMC Plant Biology[a]	6	3.44	1009
Cell Research	1	3.426	1010
Critical Reviews in Plant Sciences	3	3.4	1013
Plant Biotechnology Journal	6	3.378	1019
International Journal for Parasitology	1	3.337	1020
Plant and Cell Physiology	4	3.324	1024
Global Ecology and Biogeography	1	3.314	1025
Cytometry	2	3.293	1027
Evolution and Development	1	3.293	1028
Developmental Dynamics	1	3.169	1029
Fungal Genetics and Biology	4	3.121	1033
DNA Research	1	3.066	1034
Chromosome Research	3	3.057	1037
Metabolic Engineering	1	3.012	1038
American Journal of Botany	7	2.969	1045
Molecular Plant Pathology	13	2.963	1058
Planta	7	2.963	1065
Biochemical Society Transactions	1	2.962	1066
Analytical Biochemistry	1	2.948	1067
Heredity	2	2.872	1069
European Journal of Organic Chemistry	1	2.769	1070
Journal of Molecular Evolution	2	2.767	1072
Gene	3	2.721	1075
Theoretical and applied genetics	112	2.715	1187
Insect Biochemistry and Molecular Biology	1	2.711	1188
Rapid Communications in Mass Spectrometry	4	2.68	1192
Journal of Chromatography B:	1	2.647	1193
Molecular Genetics and Genomics	31	2.552	1224
Biotechniques	5	2.462	1229
Annals of Botany	6	2.448	1235
Biological Journal of the Linnean Society	1	2.445	1236
Phytochemistry	6	2.417	1242
Journal of Agricultural and Food Chemistry	3	2.322	1245
Journal of the Royal Statistical Society B: Statistical Methodology	1	2.315	1246
Tree Physiology	3	2.297	1249

Journal Title	Articles	Impact Factor in 2006	Cumulative Number of Publications
Biochimica et Biophysica Acta - Gene Structure and Expression	2	2.293	1251
IEEE-ACM Transactions Computational Biology and Bioinformatics,	1	2.283	1252
Mammalian Genome	2	2.279	1254
Functional Plant Biology	3	2.272	1257
Current Genetics	1	2.22	1258
Plant Pathology	1	2.198	1259
Physiologia Plantarum	2	2.169	1261
Molecular Breeding	9	2.135	1270
FEMS Microbiology Letters	2	2.068	1272
Advances in Agronomy	2	2.059	1274
Plant Biology	1	2.059	1275
OMICS	3	2.056	1278
Journal of Cereal Science	2	2.046	1280
Gene Expression Patterns	1	2.005	1281
Genetical Research	2	2	1283
Cytogenetic and Genome Research	1	1.993	1284
Journal of Animal Science	1	1.983	1285
Genome	32	1.972	1317
Genome Dynamics	1	1.956	1318
Journal of Heredity	10	1.942	1328
Molecules and Cells	1	1.872	1329
Protein Expression and Purification	1	1.867	1330
Plant Physiology and Biochemistry	2	1.847	1332
Journal of Biomedicine and Biotechnology	3	1.829	1335
Mycorrhiza	1	1.813	1336
Plant Disease	3	1.795	1339
Plant Cell Reports	1	1.727	1340
Methods in Enzymology	7	1.64	1347
Field Crops Research	2	1.634	1349
Plant Science	9	1.631	1358
Systematic Botany	1	1.629	1359
Methods in Cell Biology	1	1.582	1360
Journal of the American Society for Information Science and Technology (JASIST).	1	1.555	1361
Canadian Journal of Forest Research	1	1.549	1362
Cellulose	1	1.539	1363
Plant and Soil	2	1.495	1365

Journal Title	Articles	Impact Factor in 2006	Cumulative Number of Publications
Genetica	3	1.492	1368
Conservation Genetics	1	1.429	1369
Canadian Journal of Physiology and Pharmacology	1	1.38	1370
Annals of Applied Biology	1	1.379	1371
Comparative and Functional Genomics	6	1.341	1377
Protoplasma	1	1.333	1378
Journal of Parasitology	1	1.3	1379
Journal of Economic Entomology	1	1.275	1380
Plant Systematics and Evolution	1	1.239	1381
Advances in Botanical Research	1	1.222	1382
Canadian Journal of Botany	3	1.193	1385
Crop Science	42	1.153	1427
Parasitology Research	1	1.14	1428
Australian Journal of Agricultural Research	1	1.133	1429
Journal of the Science of Food and Agriculture	1	1.026	1430
American Journal of Enology and Viticulture	2	1.009	1432
Journal of Food Science	2	1.004	1434
Plant Breeding	3	0.954	1437
Euphytica	6	0.907	1443
Journal of Agricultural Biological and Environmental Statistics	1	0.85	1444
Journal of Phytopathology	1	0.817	1445
Journal of Nematology	5	0.771	1450
Information Software Technology	1	0.726	1451
Chinese Science Bulletin	1	0.722	1452
Nematology	2	0.722	1454
Journal of Horticultural Science and Biotechnology	1	0.719	1455
Plant Molecular Biology Reporter	5	0.625	1460
Breeding Science	1	0.622	1461
HortScience	2	0.613	1463
Genetics and Molecular Research	1	0.574	1464
Maydica	9	0.569	1473
Economic Botany	1	0.449	1474
Israel Journal of Plant Sciences	1	0.437	1475
Journal of Parallel and Distributed Computing	1	0.43	1476
Seed Science and Technology	1	0.41	1477
Brittonia	1	0.211	1478

[a]The impact factor of BMC Plant Biology was not available when this report was written. The estimated impact factor is shown.

H

NPGI-Funded K-12 Outreach Activities with Broad Potential Impact

Some NPGI-funded principal investigators (PIs) focus on providing outreach or enrichment activities specifically targeted to local K-12 students and/or teachers, but there are also a number of programs in which interested students and teachers from other parts of the state or the country can participate. Examples are listed below.

- Iowa State: Plant Genome Outreach
  - *http://www.eeob.iastate.edu/plantgenomeoutreach/7-12%20teachers.htm*
  - Seven-week, hands-on research program in molecular biology and genomics for biology teachers in grades 7 through 12.

- University of Arizona Teacher Internships in Plant Genomics
  - *http://biotech.biology.arizona.edu/RET/TIPG.html*
  - Seven- to eight-week mentored research internships in plant genomics for teacher interns to develop and share teaching materials using the tools, skills and knowledge they obtain during their research.

- South Carolina DNA Learning Center
  - *http://www.clemson.edu/cle2_share/cwe/prov5101_emphasis_areas_oview/public.www/niche.php?niche=sc_dna_learn*

Errata

Achievements of the National Plant Genome Initiative and New Horizons in Plant Biology, The National Academies Press, Washington, D.C. 2008.
ISBN-13: 978-0-309-11418-9
ISBN-10: 0-309-11418-7

On p.31, the estimated number of unique visitors and page impressions in a month for *Gramene* listed in Table 2-3 is incorrect. The numbers listed represented the number of unique visitors and of page impressions since the inception of the Website. The correct number of unique visitors in a month is 11,572 and the number of estimated number of page impressions in a month is 118,915. (Correction submitted by Lincoln Stein, Ontario Institute of Cancer Research.)

On p.31, the estimated number of unique visitors in a month for Maize Genetics and Genomics Database listed in Table 2-3 is incorrect. The number listed represented the number of total visits, and not unique visitors, in a month to the Website. The correct number of unique visitors in a month is 8,359. (Correction submitted by Carolyn Lawrence, U.S. Department of Agriculture, Agricultural Research Service.)

- o Biotechnology and genetics field trips for middle and high school students, DNA summer camp, in-service summer programs for high school teachers.

- Center for Plant Genomics Training and Education, University of California Berkeley
  - o *http://outreach.potatogenome.org*
  - o Summer research for underrepresented high school students, bioinformatics summer workshops for high school and college students

- Rice Blast Genomics Outreach
  - o *http://www.science-house.org/fungal/index.html*
  - o One-week teacher training workshops, kits, and laboratory manuals for teachers in North Carolina, Kentucky, Texas, and Arizona

- North Carolina State Summer College in Biotechnology and Life Sciences
  - o *http://www.cifr.ncsu.edu/scibls/index.htm*
  - o Laboratory courses for high school juniors and seniors for college credit.

I

Examples of Joint Call for Proposals and Co-funded Programs by IWG Members

Call for Proposal	Year	Website
The United States Rice Genome Sequencing Project; Interagency Program Announcement; Request for Proposals and Request for Input	1999, 2000, 2001	
Developing Country Collaborations in Plant Genome Research (DCC-PGR)	2004	*http://www.nsf.gov/funding/pgm_summ.jsp?pims_id=12789*
Maize Genome Sequencing Project: An NSF/DOE/USDA Joint Program	2005	*http://www.nsf.gov/pubs/2004/nsf04614/nsf04614.htm*
Memorandum of Understanding between the USDA Cooperative State Research, Education, and Extension Services and the U.S. Department of Energy Joint Genome Institute: Cooperation in Support of Coordinated Plant and Microbial Genome Sequencing and Bioinformatics	2005	*http://www.csrees.usda.gov/nea/plants/part/pbgg_part_doe.html*
Plant Feedstock Genomics for Bioenergy: A Joint Research Solicitation—USDA, DOE	2005, 2006	*http://www.science.doe.gov/grants/FAPN06-03.html* *http://www.sc.doe.gov/grants/FAPN07-03.html*

J

Examples of Interactions with Industry and Plant Breeders Reported by Principal Investigators of NPGI

A subset of the 167 responses received is listed in this appendix to show the interactions between principal investigators of the National Plant Genome Initiative and industry and plant breeders. Some responses were redacted to remove third party names.

START-UP OF BIOTECHNOLOGY COMPANIES

- Our NPGI supported work has led to the formation of a spin-off company, GrassRoots Biotechnology, which is currently in discussions with two major Ag-Biotech companies.
- Founder and current Science Advisory Board member, Divergence, Inc.
- Two start up companies were established and are still in business: Plexigen and Array Express.
- We have formed a start-up company (PhylloTech, LLC) to produce natural product fungicide alternatives and to exploit trichome produced diterpenes and sugar esters. Also, since the field trial for aphid resistance, we have had 3 additional APHIS permits (on a non-trichome project) and conducted the respective field trials.
- The technology developed at the University of Chicago on centromere biology resulted in multiple patent applications. It was used as the basis to obtain venture capital financing for Chromatin, Inc in 2001.

PATENTS OR COMMERCIAL LICENSING AGREEMENTS

- We have used a machine learning approach to identify 100 novel genes in Arabidopsis, which our analysis suggests are involved in seed development. This set of genes is being evaluated for addition to our patent portfolio by the NYU Office of Industrial Liaison. The current NYU patent portfolio on plant genes from the lab of the PI is the basis for a set of commercial licensing agreements with two major US agricultural companies working on trees, corn, alfalfa and other crop species.
- We have received continuous funding from the Biotechnology Research and Development Corporation (BRDC), a consortium of industries, to develop new plant promoters, and methodologies to improve plant transformation. We have obtained 8 patents, and applied for several others that are pending.
- Identified a soybean cyst nematode inducible promoter (patent 7,223,901)
- A co-dominant marker was developed for pungency that has been licensed to industry breeding programs.
- A segregating population has been developed for mapping that has also been licensed for private use.
- Licensed genetic stocks to Syngenta and DuPont/Pioneer

MATERIAL TRANSFER TO OR ACCESS TO RESOURCES BY INDUSTRY

- My program has provided marker assisted selection services to the following private wheat breeding companies: WestBred, Resource Seed, Inc., and Arizona Plant Breeders. Two of these companies are releasing cultivars in 2007 developed with the help of our marker assisted selection services.
- The computational tools and resources are documented at *gene.genetics.uga.edu* and *www.fgsc.net.* These websites are used to keep track of requests for software, data, and laboratory resources. Requests from companies include Pfizer, Bristol-Myers-Squib, Millenium Pharmaceuticals, MycoPharmaceuticals, Novozyme, Paradigm Genetics. Additionally, many academic labs have requested our tools.
- The funding resulted in material transfer agreements between the Regents of the University of California and Mendel Biotech
- Commercial versions of our spliced alignment software are distributed by NewLink Genetics (*http://www.linkp.com/*) and licensed to several big agro companies.
- Released open source software (TASSEL) that implements association mapping algorithms have helped companies get these to run in environments. It has been used by researcher's at all major seed companies.
- An experimental line that carries the two *G. soja* resistance QTL was released.

Seed of this experimental line was requested by 16 private soybean breeders from 11 companies and four public soybean breeders from three universities. Seed was sent to all breeders who requested the seed.

- Numerous private companies with breeding programs have accessed our web site to download our SNPs, sequence alignments and other data. These companies include Pioneer, General Mills, Monsanto, Keygene, Dow, AgReliant, Dupont and CoGentCo.
- We have produced soybean BAC contigs for the United Soybean Board, sweet orange BAC contigs for the California Citrus Board, and are in the process of generating walnut BAC contigs for the California Walnut Marketing Board.
- We have provided one seed company with gene sequences that are being used to selectively breed varieties with superior nutrition. We also trained two of their scientists in techniques in our lab. That technology is now integrated into the seed company's internal research program.
- MAGI maize genome assemblies and annotations provided to seed companies (BASF, Pioneer and Syngenta) for use on their internal maize genome web sites.
- Our software program is used by breeders at seed companies such as Seminis, De Ruiter Seeds

COLLABORATIONS WITH INDUSTRY

- One example is a successful interaction with Cotton Incorporated (Cotton, Inc. *http://www.cottoninc.com/*), which provided additional funds to generate more genomic resources for the cotton genomics community. We proposed to generate 20,000 ESTs in the NSF-funded project. Using the additional funds from Cotton Inc., we were able to produce ~40,000 ESTs. Also, we were able to design and produce the spotted oligonucleotide microarrays for the cotton research community.
- We have extensive interaction with the Phenotype Screening Corporation, which developed one of the technologies that we are using to image roots.

  (GrassRoots Biotech is incorporated in DE).
- We are assisting Cenicafe, a research institute in Colombia focused on coffee research, in identify genes important in coffee disease resistance and quality.
- We are working with industry (Bayer Bioscience and Monsanto) to coordinate the effort on sequencing cotton genomes and to transfer genomics and biotechnology to the cotton field.
- Through funding from NGPI and Monsanto company we have developed a series of near-isogenic lines each harboring a single mapped Ac transposable element. Stock requests for most lines may be made directly from the Maize Genetics Cooperation Stock Center (*http://www.maizegdb.org/cgi-bin/stockcatalog.cgi*) and

can be searched through the MaizeGDB Website (*http://www.maizegdb.org/*). These lines are freely available with no MTA required. Additional lines that we are propagating to increase seed stocks can be viewed on our project website (*http://bti.cornell.edu/Brutnell_lab2/Projects/Tagging/BMGG_pro_currentmap.html*) and seed stocks requested via email.

- We have exchanged large sets of SNPs with Pioneer Hi-Bred International. Our SNPs will be used by them for marker assisted selection and for tracking germplasm.
- We have initiated formal research/contract collaborations with Dow Agrosciences (Zionsville, IN) to help them with projects involving plant transformation and protein expression in plants.
- Collaboration with BASF Plant Sciences. The objective is to identify the Avr genes that match two late blight R genes that have entered the commercialization pipeline. The effector work is providing mode of action information as well as provide a tool for monitoring pathogen populations for virulence.
- Work built the foundation for three independent industrial collaborations on tomato flavor (with the largest European seed company), fresh lettuce producer (largest in U.S.) and a major flower seed company. Each of those collaborations resulted in a funded research project in our lab.
- Research collaboration with DuPont/Pioneer Hi-Bred to identify genes that contribute to the superior N uptake phenotype in the Illinois High Protein germplasm.
- Service agreement with Monsanto Company to conduct high-throughput amino acid analysis of plant samples.
- I interacted with Arcadia Biosciences to apply TILLING (a targeted reverse genetic technology) to wheat and soybean. My knowledge and understanding of allopolyploidy was instrumental in helping this biotech company in establishing TILLING in polyploidy species.
- This project had a close collaboration with scientists at Pioneer Hibred. The project allowed them to identify all but two of the maize cellulose synthase genes and provisional patents also resulted. Later work after this project identified the two remaining genes and also established that over-expression of certain CesA genes led to altered stalk strength, a trait important in the more windy northern parts of the U.S.; stronger stalks resist lodging in the field.

INTERACTIONS WITH BREEDERS

- Working with cotton breeders in USDA-ARS in Stoneville, Mississippi, we have successfully applied microarrays to detect gene expression differences in the breeding strains that have 20 percent better fiber strength than regular cultivars.

The gene expression data will uncover genes that are associated with cotton fiber quality.

- In addition to our basic and genomic research efforts, I run a lettuce breeding program that utilizes the genetic information on disease resistance as well as the molecular markers for selection generated by these NPGI projects. Throughout the year I meet regularly with seed industry individuals. Markers and genetic information, particularly relating to disease resistance generated by my lab are extensively used by the lettuce breeding companies.
- We are working closely with *Prunus* breeding programs worldwide to mark genes that control fruit quality characters, tree architecture and other important traits.
- We meet once a year with other soybean breeders that work in fatty composition (USB meeting) and share our discoveries. We share information so other breeders can use it to perform marker-assisted selection in their programs.
- I collaborate with the soybean breeder of the USDA-ARS unit here in NC so when I discover QTLs and develop allele specific markers for candidate genes, I transfer this knowledge to him and either he can use it to make selections or I do the marker work for him so he can make selections.
- Map and map score data for 301 recombinant inbred lines of inter-mated B73 x Mo17 (IBM), a four-fold-expanded panel of lines equivalent to over 2400 meiotic products. This resource is widely used in public and private breeding, where it provides dense and highly precise markers for marker-assisted selection.
- Breeders in India, Indonesia, Bangladesh, Korea, Ghana, Mozambique, and Uganda have all requested access to both the interspecific advanced backcross NIL families and the strategy for advanced backcross breeding in rice using wild ancestors as donors.

K

Tree Genomics or Molecular Genetics Support Provided by the U.S. Department of Agriculture Forest Service

Tree Genomics or Molecular Genetics Support Provided by the U.S. Department of Agriculture Forest Service (USFS) as Either a Competitive Award (Agenda 2020) or Cooperative Agreement from 2000 to 2005

Year	Title	Amount
	Cooperative Agreement	
2000	Host:Pathogen Signaling in Southern Pine Pathosystems	$100,000
2002	Development of a Loblolly Pine FISH-based Karyotype	$33,000
2003	Development and Comparison of FISH-based Karyotypes of Loblolly, Shortleaf, Longleaf and Slash Pines	$58,000
2005	Molecular Cytogenetics in Pine : Development comparison of FISH-based Karyotypes of Loblolly, Shortleaf, Longleaf and Slash pines	$75,000
2005	Microsatellite assessment of genetic diversity in Tsuga	$50,000
2005	Singe cell EST sequencing to define resistant and susceptible molecular host responses to hemlock woolly Adelgid	$88,000
2004	Differential gene expression in loblolly pine (*Pinus taeda* L.) challenged by the fusiform rust fungus, *Cronartium quercuum* f.sp. *fusiforme*	$44,000
	Competitive Awards	
2002	Discovery of genes controlling adaptive traits in Douglas-fir	$180,000
2003	Development of SSR markers from EST sequences in loblolly pine	$83,000
2004	Discovery of genes controlling wood property traits in Douglas-fir	$180,000
2005	Assessing the Impact of Intensive Forest Management Practices on Wood Formation and Quality at the Level of Gene Expression	$369,000

SOURCE: USFS

L

Number and Type of Mutants Distributed by the National Plant Germplasm System

Mutant Type	Number Distributed
Barley	343
Maize	3,865
Pea	31
Rice	460
Tomato	5,285
Wheat	322
Total	10,279

SOURCE: USDA Agriculture Research Service.